Artist: Don Cassity

The Quantum Bigfoot
Second Edition

By Ron Morehead

The Quantum Bigfoot

Copyright © 2017 by Ronald J. Morehead

ISBN: 978-0-9851151-5-9

All rights reserved including the right to reproduce this book or portions thereof in any form without the prior written consent of the copyright owner and the publisher of this book.

For information on Sierra Sounds, go to:

      www.bigfootsounds.com, or
      www.ronmorehead.com

5 4 3 2 1

Table of Contents

Dedication i
Statement by Bob Gimlin ii
Acknowledgements iv
Preface vi
Introduction ix
Chapter 1 : The Classical Bigfoot 1
Chapter 2 : A Biblical Perspective 15
Chapter 3 : Reported Attributes 31
Chapter 4 : Quantum Physics & Spirituality 51
Chapter 5 : Ancient Texts & Physics 65
Chapter 6 : Vibrations 75
Chapter 7 : Dark Matter 83
Chapter 8 : Language 93
Chapter 9 : Quantum Entanglement 103
Chapter 10: State of Mind 119
Chapter 11 : The Pineal Gland 125
Chapter 12 : Professional Findings 141
Chapter 13 : Quantum Time 155
Chapter 14 : Two Minds 167
Chapter 15 : Invisibility 173
Chapter 16 : UFO Connection 183
Chapter 17 : Researchers & Researching 191
Chapter 18 : Albert Einstein 197
Chapter 19 : Al Berry 203
Chapter 20 : Joan Ocean 213
Chapter 21 : Ostman's Account 221
Afterword 229
Rhonda's Sighting 232
Friends of the Project 234
Index 235

Dedication

Rhonda Morehead O'Connell

"I dedicate this book to my bigfooting sidekick & beautiful daughter, Rhonda. Her big open heart and curious mind seem to create a magnet for these beings. Whenever we are together interesting things happen in the Bigfoot realm. Thanks for all the years of research—Love you, Dad."

Statement by Bob Gimlin

March 19, 2017

It was October of 1967 when Roger {Patterson} and I saw a Bigfoot at Bluff Creek, California. It was a real shock for us when she stood up from the edge of that creek and started walking away. In all that excitement Roger was still able to get his camera out of his saddlebag and started taking a picture. He thought it would be the proof that people needed to see. What a lucky day—or so he thought. That film has become very popular, but still didn't actually prove anything to science.

It was only a few years after that I'd heard of Ron {your author} and Al Berry's recordings and their interesting story of encountering a family of these things in the Sierras. In 1978, we were all in Canada at a conference (University of British Columbia) when Dr. Kirlin gave his presentation on those sounds. Ron and I met and he invited me to visit him at his home in Merced, CA.

My plan for that summer was to ride my horse from the Mojave Desert all the way to Canada. So I loaded up my horse and mule, took off for California, but stopped and took Ron up on his invitation. Especially since I was fixin' to ride through those same mountains where they'd ran into those Bigfoots.

We visited for a while, and then I was on my way. But, I got to hear those sounds that he and Al recorded. Anyway, at that time I realized that Ron and I had something besides Bigfoot in common—it was horses, hats and bridles.

Since that time he and I have met at many of these Bigfoot meetings and became good friends. I haven't read his new book, but

I've always loved to hear his interesting speeches at these meets that we both go to. His ideas on what these beings might be are a little different than most, but then most folks haven't had the close-up, encounters that he's had.

I wish him success and hope for many more years of friendship.

Bob Gimlin

2005 Bellingham, Washington Conference.

Acknowledgements

I want to express my sincere gratitude to the three people who helped me with the editing and formatting of this book. It's a difficult task to edit a book—easy for a brain to read text differently each time. However, these three amigos did a wonderful job with my 'chain of custody' idea that we began with—back and forth and back and forth again.

Firstly, I gave the manuscript to Keri, my beloved, who looked it over and corrected issues that were obviously due to my brain working overtime. Her tireless help with the editing and encouragement is at the top of my list. She's not only my best friend, but everything any man could ask for in a partner. I relate how we met and how I feel about her in Chapter 9, Quantum Entanglement. What I didn't mention in that chapter is that we've actually known each other since she was just sixteen years old—but I was a gentleman—then.

Secondly: to my good friends, Joe and Sharon Beelart, who I've known for several years, and who worked on this project relentlessly, day after day, formatting pages and inserting pictures in the correct place and looking for errors…again and again, over and over. When it was time Sharon gave a last look and polished our layout.

Thirdly: I want to give special thanks to Molly Hart Lebherz, an amazing woman who can and did, find errors that somehow were overlooked by everyone else—it happens. We call her 'Sweet Molly'. Her heart and graciousness during this grinding editing process was, simply put, absolutely wonderful.

Fourth: My appreciation to all the physicists and professionals, who wrote articles that were absolutely essential for this book. Without the rules that apply in this world and the PhD's behind the discovery of Quantum Physics, we'd all still be in the dark about many things. I'd also be remiss if I didn't thank the artists, Don Cassity for his renderings of a good and an angry Bigfoot, and Guy Edwards for his great design for the book cover. All the artistic work is very much appreciated and the artists are acknowledged on that page.

Then we have my children, Ronika, Rhonda, Rachelle, and Royce. Throughout the years, whether on horseback, in a boat, or just trekking, they have all been very supportive of my research and were all great children, and now awesome adults. I am a very lucky man to have them in my life.

Preface

After researching the Bigfoot/Sasquatch phenomenon for over 45 years, I realized that many people are still in never-never land about these beings. Although many have their paradigm fixed, no one, in fact, knows for sure who or what these beings really are, how they originated, or how they do some of the things they do. Most of us believe almost all of what we can see with our eyes, but often doubt what cannot be seen. I believe that any mystery, through the method of science and/or technology, can be solved.

My first book, *Voices in the Wilderness* was my 40-year chronicle of how this fascinating subject started for me. Now my research points me into a direction I would have never believed in the 70s. The Quantum Bigfoot is my effort to connect all of the dots; painting a picture that will help fellow researchers and at the same time bring awareness to the public of what these beings are reportedly capable of and why. After all, if they can find us, why haven't we found them?

For me, Bigfoot is more than just a stealthy animal, like a black puma, or a Japanese Grizzly bear (declared extinct in 1964). Many such animals are very elusive and are able to stay hidden from humans. But Bigfoot is different in a way that seems to baffle scientists and researchers alike. They just don't play by the same rules.

From eyeshine-cones to tree knocks and glyphs, to how some of them might even be able to change their vibrational frequency to be out of our perception. I'll offer the science behind how I believe many of those unusual topics can be understood. It may not be the science you learned in high school, but it is an established science called quantum physics. The mathematics of quantum physics is accepted worldwide by over one million physicists. However, even with its growing popularity, many classical scientists, in a disciplined three-dimensional model, would not even discuss it with me. They want to believe that a hominid-like creature exists, but have a problem researching outside their classical disciplines.

In this book, I will quote Nobel Prize winners, physicists, and intellectual scholars. Much of the information used to support my theory was obtained from scientific articles and journals. Quantum Physics, in my opinion, is very much like spirituality and I think they must combine to get a closer look at this Bigfoot phenomenon.

Self-taught in biblical history, Greek mythology, and with an unrelenting desire to understand quantum physics, I strived to put this book together in a format that I hope will be useful to researchers and all who wish to pursue this subject. After all, most of us believe in a Supreme intelligence (God), which we can't see. We know the Cosmos is definitely out there. For many 'creation' took place in six days. But science teaches evolution is how everything actually started. Can it all mix together and make sense? For me, it can.

Ron Morehead

As you read this book you may notice that rather than using the term 'That', as in, "…these creatures 'that' roam the backwoods," I use the term 'Who', as in "…'who' roam the backwoods. This is done because I think of them as a type of sapient human hybrid, but not completely human like us. Also, in this book I've intentionally capitalized the term 'Bigfoot', as opposed to 'bigfoot' because they are not yet scientifically established as a species, e.g., bears, wolfs, humans, etc. This seems to bother some people, but I think it stands out better, so my logic has now been addressed — right or wrong; after all, I'm writing this book.

Ron Morehead, Sequim, Washington, 2017

Introduction
The Quantum Bigfoot

> "We all know that light travels faster than sound. That's why certain people appear bright until you hear them speak." Albert Einstein

It's probable that Einstein meant that quote in jest. After all, he was known to have a dry sense of humor. But there is a very serious side to the characteristic of light. Because of 'light', our eyes only observe reflections, and that observance is only available to us when light is present. Quantum physics says that 'nothing is real until it's observed'. So it appears that observation makes things real to us, but is it contingent upon having light? Or is there another way for us to see — to really see what's real? It is written that God is light (1 John 1:5).

Bigfoot is unique in many ways and I've heard many reports over the years telling of unusual, seemingly unanswerable, miracle-like, events that have taken place around these giants. I've always believed that there must be a law or rule within science behind any so-called miracle...that the paranormal is (with understanding) normal. Therefore, miracles are not miracles at all; just our lack of understanding the law that supports them. This thought pattern pointed me directly to quantum physics.

For a long time, my question was, "Can our thoughts surpass the speed of light?" Einstein said that nothing can do that, but he was referring to mass. Can anything be changed if we, within our minds, visualize and feel the circumstance differently, as Jesus taught? According to quantum physics, that answer is a resounding, "Yes."

Some want to call it pseudoscience; often it's referred to as 'new age' stuff, and sometimes its properties are denied altogether. But nevertheless, quantum physics is an established science that's at work throughout the universe and must be included as part of the nature of Bigfoot —who, believe it or not, stealthily roams this planet most often undetected.

If all of nature works within the rules of quantum physics; why not Bigfoot? We are all made of the same thing; waves and particles, energy to matter. To me, it seems ridiculous to separate ourselves from the rules that apply in the quantum world. So why not use those laws to try and understand these giants seen and described by thousands of people over thousands of years?

In this book I'm taking a layman's stab at bringing what I've learned through this science and coupling it with fascinating ancient text over the backdrop of my personal experience. I think this will offer some answers to many of the unusual reports that I've heard over my years of research.

God, Quantum Physics, Bigfoot

The intent of this book is not meant to diminish who or what God is. It is meant to help with understanding of the natural laws that He put in place when He positioned His people on earth. Many miracles are depicted in the Bible and a few seem unbelievable. Some Bible-toting, God fearing, Sunday morning go-to-church folks might be thinking they should close the book on me now—but hold on a minute—I'd like to challenge you to move forward with discernment.

I believe Quantum physics is the rule behind how a certain man called Jesus did His miracles. This would mean that God works through the laws that He created. And He often does that through His 'connected' sapient man. "The heart of the wise teaches his mouth and adds learning to his lips" (Proverbs 16:23).

For me this idealistic belief answers many questions about psychic phenomena, telepathy, biblical miracles, etc. I don't believe that all those who claim to be physic readers, telepaths, or shaman healers are malevolent. IF their core values and fundamental actions show love and caring for others, then in my mind, it comes from their God-given attribute to express that love (1 Corinthians, 13).

It's my hope that you will begin to embrace this concept because in this book I am persuaded to present a reasonable correlation between the rules of quantum science as the foundation of spirituality, and how it could relate to these creatures known as Bigfoot.

After over 45 years researching the mystery of Bigfoot, interviewing a lot of witnesses, and trying to solve the strange happenings that have taken place at our *Sierra Camp*; I think Einstein's bizarre statement, relating to quantum physics, "Spooky action at a distance" is coming into view. So goes my introduction, my creature-writing thoughts, and how it connects with spirituality and physics in *The Quantum Bigfoot*.

Artist: Don Cassity

Chapter 1
The Classical Bigfoot

For the sake of defining a giant, let's clear the air. If every one of us were 14 feet tall, we would think of an 8-foot tall person as a midget. If all of us were an average of 4 feet tall, those that were 8 feet tall would be a giant. So for clarity in this book I define a giant as 8 feet or more tall; albeit, most of us are over 4 feet tall.

So have giants really been on earth? Absolutely; the remains of giants have been found all over the world. As recently as 2013, seven were unearthed in southern Ecuador and five more just 20 miles south, close to the Peruvian border. They were between 7 and 8 feet tall; and due to their proximity may have been part of a local tribe.

I've been to Peru and Bolivia twice researching the enigmatic remains of a pre-Inca people who, according to two independent scientists, who I was with, were not completely human. My trips there were to see if the mysteries associated with them could relate to giants in North America. My crumb-trail for the connection began in South America, continued through Central America, and went all the way into Canada. But wait — back up — just how did the cartoonish name of 'Bigfoot' get its start? And could it relate to the pre-Incan people and the Ecuadorian giants?

The term "Bigfoot" was coined in 1958 by Jerry Crew for obvious reasons at the time. Jerry was operating a bulldozer for a lumber company. While clearing a road in an uninhabited and loosely surveyed territory in far northwest California he came across huge five-toed prints. The local newspaper got wind of this and folks were alerted — something with a big foot is roaming the mountains of Humboldt and Del Norte counties in Northern California!

The Classical Bigfoot

All over the world, it seems *beings* like Jerry Crew's northwest California "Bigfoot," or similar creatures, have a name tagged to them like Sasquatch, Wildman, Yeti, and Yeren, being well-known examples. Whatever they are called, there is a mystery as to what they actually are. Did they evolve naturally, as science says we did? Do they all share the same genome? Can they cross-breed with humans? Could they be of alien origin?

Most of the scientists who privately believe these creatures exist think there could only be one type of species and/or sub-species. Therefore their anatomy must be the same or at least very similar. This suggests to me that those scientists are not really considering other possibilities, but they have acquired the discipline that requires them to examine the facts and nothing but the facts.

The commonalities of these giants are they are very hairy, have very human-like facial features, and can move very fast. The Bigfoot family who I interacted with, and will describe later, has a complex language. Does this mean that all of them use and have language? In John Green's interview with Albert Ostman (purported Bigfoot abductee), he said that they were chattering with some kind of language (Chapter 21).

Some professionals say that if one has the attribute for language they must all have that attribute. Some witnesses report a sagittal crest while others say, "No, the head was rounded like humans." Their footprints seem to vary profusely, but due to the Patterson film of 1967, scientists want to judge all prints by the 'for sure' prints cast at that film site. However, as noted in the different pictures on pages 3 and 4 all our prints from the Sierra camp area, whether in the snow, at camp, or on the ridge line, were very splayed and had very little arch.

The Classical Bigfoot

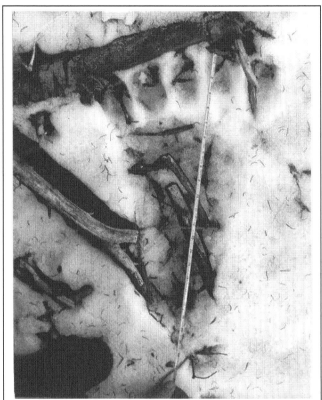

Late fall 1972: Track found within 20 feet of our shelter. Length slightly less than 20 inches.

The Classical Bigfoot

Plaster casts and foot impressions from N. California

The Classical Bigfoot

There is a large group of people who think they must be a remnant of the gigantopithecus, a giant ape-like creature who supposedly went extinct 100,000 or more years ago. Gigantopithecus was discovered in 1935 by anthropologist, Ralph von Koenigswald who found fossilized molars of a genus of giant prehistoric ape in a Hong Kong apothecary shop. After a lot of head-scratching and hypothesizing, this amazing creature came to be called gigantopithecus.

The few scientists who are looking for an answer to the "giant" enigma usually turn to the gigantopithecus. This is because more fossil remains of gigantopithecus have been found since Koeningswald's discovery. With over 1,300 teeth, mandibles, and fossilized bones from three established species of the giant ape, scientists have something they can analyze.

Unfortunately, suitable remains have not been discovered to get reliable DNA (deoxyribonucleic acid). From the remains found we know a male adult gigantopithecus was probably between ten and twelve feet tall. However, we don't know if they walked upright, or were quadrupeds like existing apes.

If gigantopithecus evolved, like evolutionists say humans evolved, there would be no more of a spiritual significance in them than in any other wild animal. Furthermore, if Bigfoot is somehow a remnant of the gigantopithecus, he may only be a giant ape who has learned how to live undetected in the woods with no unusual attributes at all. However, logically I must ask, "With all the people looking for the being, and all the technology available, how does something so big remain so hidden for so long?"

Maybe a Hybrid?

"Stuart Fleischmann, Assistant Professor of Comparative Genomics at the Swiss University in Cairo, and his team, have published the results of a seven year study mapping the genomes of nine ancient Egyptian Pharaohs. If proven correct, their findings could potentially change the world's history books forever, and our understanding of our place in the Universe!

The most exciting thing had happened after initial tests took place. Eight out of nine samples returned with rather interesting but typical results. The ninth sample belonged to Akhenaten, an enigmatic 14th century BC pharaoh, and father of Tutankhamun. It was a small fragment of desiccated brain tissue which had been the source of Akhenaten's DNA sample. The DNA test was then repeated using bone tissue — and the same results were obtained. None of the results came back as a completely normal human.

Now this is where it gets very interesting indeed, it appears this increased activity in Akhenaten's genome would suggest he had a higher cranial capacity because of the need to house a far larger cortex. But some ask, wouldn't a mutation have caused a human brain to grow like this? Well as of now we have yet to discover such an intriguing technique, which is despite many years of breakthroughs in genetic science. Could it possibly be that this over 3,300-year-old evidence points towards ancient genetic manipulation on a human subject? Is it potentially the work of advanced extraterrestrial beings?"
Source: UFO International: .ufointernationalproject.com

Separated by an Ocean

Separated by the Atlantic Ocean, Akhenaten's skull and the skulls we examined in Peru and Bolivia seem to have a common cranial elongation. In addition, according to Dr. Aaron Judkins, Biblical archeologist, the elongated South American skulls were lacking a sagittal suture and after weighing them we found they would have had more brain matter, weighing 25% to 30% more. Dr. Judkins also determined the Peru and Bolivian skulls were not completely human.

Who knows where Bigfoots truly came from, but nowadays they are alive and seem to be doing very well; a huge bipedal ape-like being, who roams the earth at will, stays hidden in the backwoods and rarely interacts with humans. But this is where my theory separates from the tribe and begins to circle a different scientific wagon train.

In the case of the Peruvian naturally elongated skull on the next page, many of these types of remains are displayed in Paracas, Peru—mostly in private museums. This example, as with many others, is not a form of Cranial Deformation. This hybrid skull has no sagittal suture (only one parietal bone), unlike human skulls which have two.

A Comparison of Ancient Egyptian and Peruvian Skulls

The Pharaoh Akhenaten 1350 BC

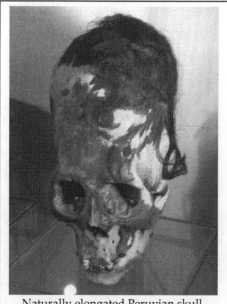

Naturally elongated Peruvian skull.

The Classical Bigfoot

Obviously, like many, I think they are much more than just an ape-like creature living in the woods. Based on my experiences around them they are definitely cognizant, and from a highly qualified Crypto-Linguist, primarily responsible for identifying foreign communications using signals equipment, the recordings we made of these beings in the Sierra Camp have a complex language. They reasoned, not like a wild animal, but like a human with sapience, that is, a human who is wise and discerning. Does this mean that all sapient beings are human?

> "A human being is part of the whole called by us 'Universe...a part limited in time and space. He experiences his thoughts and feelings as separate from the rest — a kind of optical delusion of his own consciousness. This delusion is a prison for us restricting us to our personal desires and to affection for a few persons nearest to us. Our task must be to free ourselves from this prison by widening our compassion to embrace all living creatures and the whole of nature and its beauty."
> Albert Einstein

If Einstein said we should be compassionate of all living creatures does that mean that all Bigfoots deserve our compassion — good or bad? I think the family of these creatures that I've dealt with in the Sierra Nevada Mountains was friendly; they trusted the environment of our remote camp. Yet they never came out from behind huge trees and purposefully allowed themselves to be exposed. Only on occasions did any of us catch glimpses.

The Classical Bigfoot

At first, we thought their vocalizations were directed at us and obviously sounded very threatening, but in hindsight, those sounds may have been directed at one of their own — someone else wanted the leftover food, or maybe getting too close to these little sock-headed hominids inside that shelter. They only took food we left for them. They never took the deer we hunted which was hung on a cable behind our shelter, although at times they did move those carcasses around a little.

Over the years I've heard several accounts of these creatures being aggressive toward humans, sometimes throwing rocks. My opinion is that rock throwing was due to humans going into an area where the Bigfoot didn't want them to be — that seems reasonable, possibly protecting a young one. Or, perhaps they wanted to hunt where the human was. That really makes sense; we all have to eat. But maybe the best one is that they were having a bad-hair day, or maybe during a nap one slept on a prickly pinecone — who actually knows why they scare people off at times by throwing rocks? It's all speculation.

The Classical Bigfoot

Sierra Camp: Bill McDowell (seated) and Louis Johnson
Photo by Ron Morehead

Why they interacted so much with us at the Sierra Camp is a question still being asked today, and the answer I used to give is that they were curious and loved our food. However, I now think there may be a more complex reason, a spiritual reason, for this to have happened. Speaking on behalf of the good Bigfoots; they want us to understand more about them, or better yet, more about how humanity is screwing things up — yet they may be forbidden to actually interfere. If that is the case, just who, or what, could have forbidden them?

A Case of Inspecting, Observing, & Non-interference

Adapted from Chapter 6, *Voices in the Wilderness*, Morehead, 2012. (Note picture on opposite page).

"Bear ravaged the deer if we left them in the forest overnight, so often it was a challenge to beat the fading daylight. Sometimes Bill and I needed to make more than one trip to pick up all the deer. More than once, when we arrived back in camp to get our horses, we found big five-toed footprints around them. Ah, if our animals could only talk, or operate a camera!

After our return to camp, we'd hang the deer on a cable behind the shelter where later they would be skinned. Sometimes, after a visit from our giant friends, the next morning we noticed huge footprints under the deer. Figuring they were curious about the skinned deer, late one afternoon, Larry, Warren's son, and I smoothed out an area under the deer – hoping to obtain a really clear print. We then joined the others around the stove for the usual feast of fresh liver and heart.

About an hour had passed when Larry and I took our flashlights to check the backside of the shelter again. The deer had been moved down the cable and there were three big foot impressions on the ground. The shelter is about 8 feet high, and the stove is about 50 feet from where we hung the deer. Although I figured these creatures had been observing us all along, this real-time spy-job really brought that awareness to the forefront of my mind — it was a bit unsettling."

The Classical Bigfoot

Behind our Shelter

Sierra Camp shelter

Where do we go from here?

In this universe, there must be an innumerable amount of different advanced races of beings, some good, some bad, some probably very ugly — in our eyes anyway. Different ones have visited earth and mingled with the different creatures here, including humans. But after decades of research and contemplation, I believe some mingled for profound spiritual reasons.

Science ascribes to evolution, the doctrine of Charles Darwin (1809–1882), Darwin published his theory of evolution with compelling evidence in his 1859 book, *On the Origin of Species*. Darwinists would most likely pick the gigantopithecus theory. However, if you are an agnostic, a spiritual person, or maybe even semi-religious, this next chapter will surely peak your interest.

Chapter 2
A Biblical Perspective

Ancient Ones, Aliens, or Hybrids?

To get an idea of what the ancient texts say about giants we should look at those texts and see if any could relate to our current day Bigfoot. Greek mythology and biblical accounts state that some demi-gods are a product of aliens copulating with humans. Those hybrids, by their creator's clever manipulation of the DNA, were given advanced attributes. Those attributes could entail the ability to work within the quantum level of consciousness — just as humans, in my opinion, were originally designed to do.

Because of my on-going close-up experiences with a family of Bigfoot, I'm often asked, "What are they? Are they good, or bad?" My answer was/is that I really don't know. The ones that I and my hunting buddies encountered in the high country of the Sierras Mountains of California didn't harm us. We were not devoured, nor had we been carried away in our sleeping bags to be used for meals, mating, manipulation, or, who knows?

Over time, we actually thought they were becoming friendly. I've often wondered if they were trying to tell us something about themselves, or from a completely different side of the coin, they might have been actually bad; trying to gain our trust and then trick us to the dark side.

That said, I've heard many reports of these creatures carrying on maliciously and antagonizing people. So, like people, they can be nice or not nice. Right? A deeper, more profound thought is that Bigfoot have different genomes, are a product of a type of celestial being and have a completely different agenda toward humanity.

Although hybrids normally cannot procreate, a prime example of an exception today is the mule, the product of a female horse and a male donkey. It is extremely rare for mules to reproduce but it does occasionally happen. Could these beings known as Bigfoot or Sasquatch be a hybrid, who can successfully reproduce?

Humans have not been shortchanged with any of the attributes that aliens may have passed on to their creations. According to Jesus, we can do everything He did and more (John 14:12). He knew and used the laws that God put into existence — in particular the laws of quantum physics. So why are we not walking on water? Whatever these creatures are, they are not like us and have not been given the same rights that humans have on this earth.

Ancient Ones

For years I've heard of Bigfoots being referred to as Forest People, who want to protect the earth. They are also known as the Ancient Ones. These, from a biblical standpoint, would not necessarily have had their genome corrupted by the Dark Side and are considered by many to be helpful to us humans.

They might also consider them from the lineage of Cain, the first born son of a sapient man (Adam) and a sapient woman (Eve). Cain turned out to be a very bad guy and killed his brother Able. He was 'Marked' by God, but still lived on, got married, had many children, and did malicious things on earth; as did his many warring decedents. There is a very conservative estimate of seven million people on earth prior to the flood of Noah. Some have estimated there were many more than that.

It's not unreasonable to believe that one of Cain's offspring could have been the wife of Ham, one of Noah's sons and builder of

A Biblical Perspective

the ark which housed Noah's family during the flood. This would mean that Cain's lineage, via Ham's wife, made it through the flood and their descendants were not the Nephilim; the hybrid demigod giants from celestial intervention into humans.

In Genesis 4:23, prior to the flood, we find that Lamech, Cain's great-great-great-grandson was an awesome intuitive hunter, but blind. He was best known for pursuing game blind, but with his son's help. During a hunting trip Tybal-Cain, his son, thought he had identified a wild animal for his father to shoot, but it was Cain and Lamech inadvertently killed him.

So, this means Cain had the look of a wild animal; obviously a different look than what they had, which also means that the 'Mark of Cain' was not a mutation of his DNA, or it would have been passed on to his descendants. However, this story gives me pause. How could anyone tell a blind person where to shoot, trying to direct him to a target?

My thought is that maybe some of the original attributes given to sapient man could still be understood and easily accessible, e.g., intuitiveness, telepathy, a Sixth Sense. So through use of a Sixth Sense, perhaps Lamech 'sensed' his targets. See more about this attribute in Chapter 11, "The Pineal Gland."

The mingling of the sons of Adam is reasonable and would lend credence to the notion that some Bigfoots may not be a remnant of the Nephilim, and the lineage of Cain could have been aware of the attributes that mankind was originally given. However, for killing Cain the curse that transcended to Lamech was seventy and sevenfold. What could that mean? Ham was one of Noah's three sons. Could whatever that curse was have been passed on through Ham's wife?

A Biblical Perspective

Ham's decedents ended up inhabiting the land of Canaan where the giant Goliath and his brothers lived, but the Nephilim were also in that land (Numbers 13: 33-34). The Bible doesn't refer to Goliath as a Nephilim, but he may have been. He was certainly a huge man, as were his three brothers.

Considering giants were in Canaan hundreds of years after the flood it's very believable that there was a lot of copulating and mixing it up happening there. My explanation of how the Nephilim, who were once destroyed by the flood, inhabited Canaan after the flood, is simple: If entities from the cosmos with free will did it once in pre-flood days, others could have done it again by co-mingling with the Canaanites, who were historically a warring people, thus creating giant hybrids in another attempt to corrupt the human genome.

Many who think the Bigfoots want to be our friends and are not wicked could hold to the belief that the decedents of Cain were not the Nephilim, they were more human-like, with human attributes originally given to man. Thus they were the true Ancient Ones. However, they inhabited the land of Canaan where a negative culture existed, but some of Cain's uncorrupted lineage may have fled prior to the Israelites overtaking that land.

The Book of Enoch, discovered in Ethiopia in 1768, goes into great detail about the Fallen Ones, the celestial beings who went against God's will. In 1956 multiple copies of that book were found with the Dead Sea Scrolls.

In the book of Enoch, we find detailed accounts of giants and what was happening during those pre-flood days.

When the Israelites invaded Canaan, God gave instructions to kill all the people; men, women, children, even the animals. Really, even the children? One might ask, "How could the God of love be so cruel?" According to the Bible it was because the land of Canaan was corrupt, had been penetrated by the dark side, trying desperately again to stop the messiah, who had been prophesied to come from a pure 100% human genome.

An Alien Hybrid

But here's another biblical idea, i.e., Rebekah's womb. She had twins, one covered with red hair (Esau) and one normal (Jacob). In another shrewd and sneaky way, the dark side may have still been working the angles. Now the big question; can two eggs in a woman's uterus be fertilized by two different males at the same time? The answer is 'Yes'. One in 400 sets of fraternal human twins is bi-paternal. (www.babycenter.com).

Advanced beings can manipulate, through quantum rules, the human makeup and may have fertilized one of the two eggs in Rebekah's womb thus trying one more time, in a completely different way, to corrupt the human genome. All the birth rights would have been given to Esau, being the first out of the womb. The dark side was not giving up; just being more cunning.

Through trickery, Rebecca instigated a method by which Jacob, instead of Esau, received the blessing from his father and thus received all the birth rights (the right to inherit the father's estate). That lineage of Esau (her first born twin) became the father of the Edomites (Edom means Red) who inhabited northern Canaan. Many mingled and spread into different cultures and races; and, as many as 10,000 were slaughtered by Amaziah (2 Chronicles 25:14).

A Biblical Perspective

Edomites today are considered a mindset instead of a race of people; very undesirable beings. It's been suggested that the cave pictured at the bottom of Michelangelo's painting of the Last Judgement (p.29) is a depiction of hell, and inside that cave are two Edomites.

Government Involvement?

But there is possibly another, more disturbing type of DNA manipulation in primates, not biblical at all. That is government intervention and experimentation. Prior to WII Russian dictator Stalin searched for ways to naturally produce "super warriors." Adolph Hitler's henchmen also worked on the "super warrior" idea during World War II. Some believe that our government picked up the Hitler wand and has carried it to a new level; possibly an out-of-control level. If there's anything to this theory we should all be on our knees; really, the United States government?

Missing 411

Although David Paulides is not discussed at length in this book, I'd like to note he has written books about missing people, mostly in State and National Parks, who have never been accounted for by our government. I've been at meets with Mr. Paulides and his book series, Missing 411, is to be complimented.

David's background as a professional investigator has compelled him to give the facts, and only the facts. He writes about numerous people who went missing and the government doesn't know what happened, or they're simply not telling us. Anything else is speculation.

That said, we will probably never know if those missing people just wandered off a cliff and a bear ate them, or if giants took them and did something sinister...most of them and/or their remains have never been found. Native American lore speaks of this type of behavior regarding the Bigfoot/Sasquatch.

Alien?

The last theory is that they are coming from another place in the cosmos and enter earth through portals. So, there may really be a UFO connection. Several reports of UFOs are said to also have a Bigfoot connection, (See Chapter 16, A UFO Connection?).

How do some of these ideas lend themselves to Quantum Physics? First, Noah was a spiritual man and knew the quantum attributes that were originally part of mans' awareness. That knowledge was likely passed on to his descendants.

Nimrod, Ham's grandson, who has been credited for attempting to build the Tower of Babel (Genesis 11:1) was a descendant of Cain, and he knew the unique powers given to man. He was a hunter and a vicious warrior.

If we take into consideration that men still knew that within them the quantum power to create could be solidified by their words then the Tower of Babel would have made sense to them; speak of it and it can happen (Genesis 11:6). The Bible says that their common language was broken up. Thus, all the collective energy of their 'words' was taken away.

In my opinion, some of these reclusive beings (the bad ones) are a product of a cross between disobedient humans and aliens, e.g., the Nephilim. Genesis 1:26-27 states clearly how we were originally designed:

A Biblical Perspective

> And God said, "Let us make man in our image, after our likeness; and let them have dominion over the fish of the sea, and over the fowl of the air, and over the cattle, and over all the earth, and over every creeping thing that creepeth upon the earth."

To sum this up; if some Bigfoots are the Nephilim, they would most likely understand and operated in the quantum realm. If they are a descendant of Cain, they could be the Ancient Ones and want to help mankind in hopes of having a place in the hereafter; still understanding, and using, the attributes that God's 'man' originally had access to. If they are a derivative of our government messing around with a super warrior idea, we could all be in big trouble. If, on the other hand, they are a product of evolution and just a very stealthy ape in the woods, why haven't we been able to find one?

A Cosmic War on Earth

Bigfoot artwork by Don Cassity, layout by author

Humans have guardian angels and we could entertain them, unaware (Hebrews 13:2). From the beginning, we're told in the scriptures who and what we are. In 1 Peter 2:9 we're told about our future as God's people. How arrogant of us if we belittle this truth and undermine our potential?

Part of our human makeup, and who we were made to be like, became recessive in us via a negative influence with the first sapient man that God completed. We were spiritually separated from our creator in the Garden of Eden because of that metaphoric apple. The awareness of that disconnecting aspect was brought back to us over 2,000 years ago and it was clearly explained how we could reconnect with what God's man originally lost.

Jesus was called the second Adam. Could it be that Christ was the reincarnation of Adam and what happened at Calvary corrected that karma?

A Biblical Perspective

Bigfoots can be either good or bad, depending on who their creator was and their intent. We must use discernment when dealing with them. There's definitively nothing to be afraid of if you've got all your spiritual armor on. Know just who you are as a human. When the end of days comes is it possible that Bigfoot can be entangled, quantumly, by their creator to do whatever is commanded. There is a cosmic war between good and evil happening now.

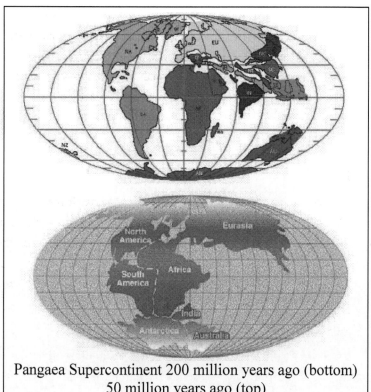

Pangaea Supercontinent 200 million years ago (bottom)
50 million years ago (top)

A Biblical Perspective

Many geologists from around the world agree that there was an earth-shattering event between 14, 000 years BC and 11,000 years BC when landscapes were completely changed. But many Christians, me being one, were told that the earth is only 6,000 years old. What about that? I have always questioned the timelines many scholars ascribe to — just never made sense to me.

Bishop James Ussher's (1581 – 1656) chronology of the biblical timeline of creation seems to be flawed. However, this has been the timeline most Christian religions have been using for centuries. Bishop Ussher also predicted the end of the earth was to be in November of 1996. My suggested read is an interesting analysis on a different time-line by Dr. Hugh Henry, Ph.D., and Professor Daniel J. Dyke, MDiv, Mth.
http://www.reasons.org/articles/from-noah-to-abraham-to-moses-proof-of-genealogical-gaps-in-mosaic-literature-part-1

> Luke 17:26..."And as it was in the Days of Noah, so shall it be also in the Days of the Son of Man."

We now live in a time that is very similar to how it was prior to The Flood. And, it should be important to be acquainted with what it was like on earth prior to that great flood, when giants were on the earth. That flood was a monumental deluge like none ever before or since. Science has suggested how at one time the continents were torn apart by this catastrophe and from a world map, it appears very believable.

A Biblical Perspective

In Puma Punku, Bolivia, near Lake Titicaca, there is visible evidence of a massive deluge — yet Puma Punku is over 12,500 thousand feet in elevation. It seems as though the continents were raised. Could this catastrophic event be the flood that has been mentioned in many cultures and religions worldwide? This natural disaster was supposedly brought about because God's man failed, the corruption had almost taken over completely, and the giant cannibalistic Nephilim were devouring all.

Ruins at Puma Punku, Bolivia. Notice straight cuts, and especially the size of the large square slab right-center.
Photo by Ron Morehead

Cannibal Giants: Nevada

For eons, the Paiute Indians have had a story about cannibalistic red-haired giants who were eating the Paiute people. That story was established as credible when giant human-like remains were reportedly discovered in 1911 by bat guano miners in the Lovelock Cave of Nevada. According to legend, after a coalition of Natives from around the area set fire to that cave, burning the giants out, they took the cave over and used it for themselves.

Author at Lovelock Cave (4th expedition). The cave entrance is the dark shadow top center. It was near the edge of an ancient lake now dry.

The cave is located about 20 miles outside of Lovelock, Nevada. Remains of Paiute Indians and their artifacts are displayed in the Winnemucca museum, but remains of giants formerly on display have been taken away.

A Biblical Perspective

The museum and the Bureau of Land Management both deny there were any giant remains found in that cave; "just robust people", they say. However, old, and recent photographs suggest something different.

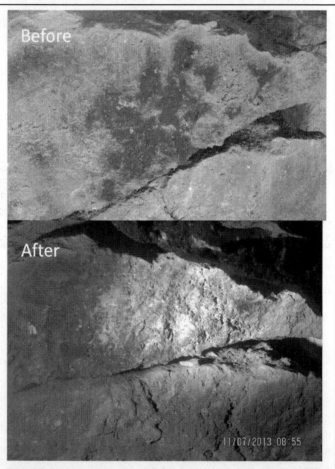

Possible huge and ancient hand print we found in the Lovelock Cave. After publicly announcing the find, soon afterward someone removed it. Who and why? An archaeological site requires a permit.

Photos by Ron Morehead

A Biblical Perspective

Michelangelo's *The Last Judgment*
Edomites in the cave

Chapter 3
Reported Attributes

Vision, Smell, Speed, Weight, Glyphs, Portals.
Tree-knocks — What's Real, What's Knock, and What's Pop?
What about Linguistic Abilities?
Can they Mind-Speak? Can they disappear?

Many of these creatures seem to have unusual attributes that are associated with them — at least this has been a part of several reports. Many of the mysteries are surely answerable using classical science, ones such as how do they see so well in the dark, how do they move so fast and effortlessly, etc.? All this should be thoroughly explored with all the classical means possible prior to stretching into the quantum field.

Eyeshine and Night Vision: Bigfoot can see well in the dark because their eyes are bigger and can collect more starlight.

> "The eyes' light gathering power goes up with the square of the diameter, so a 10mm pupil, a very reasonable size, would give Bigfoot four times the night vision capabilities as a normal human."

Dr. Henner Fahrenbach, PhD, Laboratory of Microscopy, Oregon Regional Primate Center.

Lawrence Leavell: "The most important night vision adaptation seems to be the tapetum lucidum, or simply tapetum. The tapetum is a reflective layer behind the retina, which reflects light forward through the retina, providing the retina much more light on the second pass.

But there is more. This is where the lensing structure is somewhat different. The remaining reflected light can then be transmitted out of the eye through the eyes' lens system as a concentrated "cone" of light to further illuminate the scene in the center of the eye's field of view, providing even more light upon its return once more, this time concentrated in the central region of the retina.

Some individuals, usually when young, have better night vision. It's been noted that faint perceptible "cones" of light from the eyes of cats and other nocturnal animals has been witnessed. "

An eyewitness account; watching a Bigfoot:

(2008) "....what I witnessed at the 5-6 ft. mark was a tale straight out of science fiction. In a head-down posture, the bipedal closed the gap; I became aware that it may not have known I was there. As it no longer came directly at me but paralleled my front as it noted the peanuts on the ground. At that very moment, a gradually intensifying set of small white almost searchlight or spotlight quality "cones" of light illuminated the ground at its feet emanating from its eyes with its head down as if it was concerned about the peanut trail.

At the 20-30 second count, the white cones of light dimmed out gradually as suddenly as they had come on. I was really nervous at this point because I no longer could be assured I was dealing with a variant Bigfoot (and to this day am not totally satisfied I was); but think the best explanation is the simplest one, which is yes, there is a variety with very unusual eye function." www.bfro.net

"Cone cells, or cones, are one of three types of photoreceptor cells in the retina of the eye. They are responsible for color vision and function best in relatively bright light, as opposed to rod cells, which work better in dim light. Cone cells are densely packed in the fovea centralis, a 0.3 mm diameter rod-free area with very thin, densely packed cones which quickly reduce in number towards the periphery of the retina. There are about six to seven million cones in a human eye and are most concentrated towards the macula.

A commonly cited figure of six million in the human eye was found by Osterberg in 1935. Oyster's textbook (1999) cites work by Curcio et al. (1990) indicating an average close to 4.5 million cone cells and 90 million rod cells in the human retina.

Cones are less sensitive to light than the rod cells in the retina (which support vision at low light levels), but allow the perception of color. They are also able to perceive finer detail and more rapid changes in images because their response times to stimuli are faster than those of rods.

Cones are normally one of the three types, each with different pigment, namely: S-cones, M-cones, and L-cones. Each cone is, therefore, sensitive to visible wavelengths of light that correspond to short-wavelength, medium-wavelength, and long-wavelength light. Because humans usually have three kinds of cones with different photopsins, which have different response curves and thus respond to variation in color in different ways, we have trichromatic vision.

Being color blind can change this, and there have been some verified reports of people with four or more types of

cones, giving them tetrachromatic vision. The three pigments responsible for detecting light have been shown to vary in their exact chemical composition due to genetic mutation; different individuals will have cones with different color sensitivity. Destruction of the cone cells from disease would result in blindness."
https://en.wikipedia.org/wiki/Cone_cell

Eyes that glow in the pitch-black night make for many a scary tale. But why do some animals' eyes glow at night?

"A lot of the animals that we see, especially the ones that go out at night, have a special, reflective surface right behind their retinas," says Dr. Cynthia Powell, a veterinary ophthalmologist at Colorado State University. That light-reflecting surface, called the tapetum lucidum, helps animals see better in the dark.

When light enters the eye, it's supposed to hit a photoreceptor that transmits the information to the brain, Powell explains. But sometimes the light doesn't hit the photoreceptor, so the tapetum lucidum acts as a mirror to bounce it back for a second chance.

A large number of animals have the tapetum lucidum, including deer, dogs, cats, cattle, horses, and ferrets. Humans don't, and neither do some other primates. Squirrels, kangaroos, and pigs don't have the tapeta, either. Dr. Cynthia Powell, (a veterinary ophthalmologist at Colorado State University).

npr.org/templates/story/story.php?storyId=96414364 October 2008.

Reported Attributes

Not all animals' eyes glow the same color. Powell says this is due to different substances — like riboflavin or zinc — in an animal's tapetum. Also, she says, "There are varying amounts of pigment within the retina, and that can affect the color. Age and other factors also can change the color, so even two dogs of the same species could have eyes that glow different colors.

Cats often have eyes that glow bright green, though Siamese cats' eyes often glow bright yellow. Cat tapeta also tend to reflect a little bit more than dogs."

Olfactory Sense:

Not to come across trivial, but for a long time I didn't know how to explain the difference between 'smell' and 'smell'. Olfactory is an impressive word for how well something, or someone, can pick up a scent (not how bad or how good their mass might smell), but how attuned their nose is to the environment. Bigfoot is a primate, and as a primate, his olfactory sense would not be acute...not even close to a dog or a bear. However, having not been contaminated with the by-products of civilization, it is likely much better than ours. Their wide nose, which is often reported, would allow them to take in more oxygen as well, which might be advantageous for their reported high-speed movement.

How Fast Are They? Fast, really fast. My first glimpse of one of these creatures was the evening when I was recording an exchange that my friend and I were having with them at our Sierra Camp. The one I saw moved so rapidly, through the woods, without even a little bobble, at an unbelievable speed. Since then, I've heard several accounts of how fast they run. Some reports have had them keeping up with their car at 50 mph. A normal horse runs at around 30 mph, but racehorses have been clocked at 44 mph. So as unplausible as it may seem — I'm going to believe it. If they are as big as their footprints and stride suggest, their legs, at least the bigger ones, certainly have longer legs than many horses.

Associated Sounds: Besides the rapid chattering among themselves, heard by some humans, they will clack rocks together (rhythmically). They will do the same thing with wood-on-wood, or tree knocks. I've heard one of them clacking rocks and another answering with wood knocks. I personally believe this is their method of signaling each other and perhaps signaling a way for a specific action...a code for messaging back and forth. In 2016, during a trip into the Canadian wilderness, the First Nations guide who I was with said they believe the wood knocking was territorial and when it's heard, it means for them to retreat, so they don't go any further.

They also have different types of 'whooping' sounds. I believe each type of whoop may represent something different. All of this might be sending a specific message to contact one another for positioning, warning, or perhaps it could be to see a human's response. The whoops could also be mistaken for an owl...the amplitude, however, should alert the researcher to either look for a big owl or possibly a Bigfoot.

Christopher Murphy put together an outstanding book, *Know the Sasquatch/Bigfoot*, and in that book, he gives the account of Albert Ostman from 1924. Ostman said that he was taken to Toba Inlet, BC Canada, by an old Indian and dropped off. He was looking for gold, hiked for a few days, then one night he was carried away in his sleeping bag for several hours over rough terrain. He said that he was held captive by a family of Sasquatch for six days. John Green's book, Sasquatch, the Apes Among Us, has a detailed account of Ostman's captivity. Ostman said, "...I heard them chatter — some kind of talk I did not understand." (For more on Ostman; and my expedition to assess his story see Chapter 21.)

Vocal Ability: The one thing I have difficulty getting across in my talks is how huge their amplitude can be — it can really jar a person, me being one. As shown on the two graphs that follow, their vocal mechanism exceeds what the average human can do. Their vocal range can be above, in, and below, the human range This ability suggests they are or can be, great mimics of other animals...including humans. Their vocal mechanism is very flexible. (See Chapter 12: Professional Findings: Nancy Logan).

On the next page are graphs comparing Dr. Kirlin's' findings on Bigfoot pitch and vocal track lengths to humans. Please note Dr. Kirlin's findings clearly demonstrate the *Sierra Sounds* recordings mostly fall outside the 95th percentile of normal human male speech illustrated in the boxes.

On several occasions, I've had witnesses tell me they've heard their name being called — or their dog's name when no other human was around. These statements are not unusual. According to Scott Nelson, Crypto-Linguist, they have a complex language of their own. (Again, see Chapter 12.) They communicate vocally to each other and have been known to try and communicate with humans (Sierra Sounds).

Reported Attributes

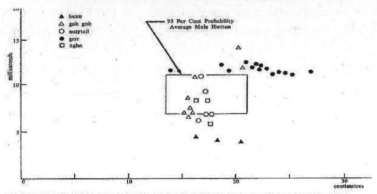

FIGURE 4. PITCH AND VOCAL TRACT LENGTH ESTIMATES WITH APPROXIMATE 95 PER CENT PROBABILITY REGION FOR NORMAL HUMAN MALE SUPERIMPOSED

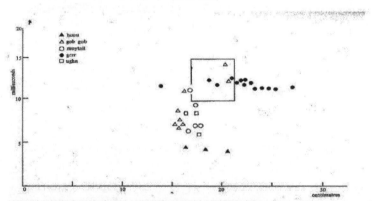

FIGURE 5: SAME DATA AS IN FIGURE 4, BUT 95 PER CENT PROBABILITY CORRESPONDS TO LOW-PITCHED HUMAN MALE AND THE VOWEL ɔ WHICH REQUIRES A LONG TRACT LENGTH

Reported Attributes

How Big Do They Get? In snow at Sierra Camp, we had small 9 inch prints alongside 18-inch prints. Accordingly, when one was spotted walking by our shelter, it was reported to be over 8 feet tall and it had an 18-inch print. That said, and away from our shelter area, we stumbled across four huge 25 ½ inch prints (one pictured below) that had a distance between them of 13 feet — results by a variety of calculation methods suggest those prints were made by a Bigfoot at least 12 feet tall. There was no indication from the prints that it may have been running.

If I hadn't been there and seen it with my own eyes, I'm not sure I would have believed it to be real. We all kept quite about that one for a few years. But since then I've heard reports of people who claim they saw King Kong, "It had to be 12 foot tall" they said.

25 ½ inch track from the following trackline photograph

Reported Attributes

Track line near *Sierra Camp* of four 25 inch tracks with a 13 foot distance between them. Donald Hilton stands next to a track. Bill McDowell measures another. 1972 photo by Ron Morehead.

Reported Attributes

When interviewing two seemingly credible men in Oregon, they told me they'd seen King Kong, "It had to be 12 foot tall", they said. Two other hunters from a completely different area in California witnessed one for several minutes and said it had to be at least 10 feet tall.

Why is it that a lot of folks can believe a creature 8 feet tall, but 12 feet tall is — well, unbelievable, inconceivable, and unfathomable. Eight feet tall or 12 feet tall, being so big, you'd think these 'Whatevers' would have been scientifically exposed by now.

When Albert Ostman gave his interview to John Green he estimated the weight of the older female, which was over seven feet tall, to be between 500-600 pounds. The weight of muscle mass is heavier because of the denseness of the tissue, and it's very rare to hear a report about an obese Bigfoot...actually, I never have, very muscular reports are most common. They can leave very deep impressions in the ground. After visiting a recent sighting, I couldn't even make an impression in the same turf.

Scott Nelson, Rhonda Morehead O'Connell and Wendy Burr at a trackline near Sierra Camp after a sighting.

Tree Knocks

I've never heard of anyone who has ever seen a Bigfoot whacking on a tree, i.e., Tree Knock, yet many folks, me being one, have heard that type of sound, many, many times and assume it's someone, or something with an opposable thumb, whacking on a tree. So, are these 'beings' trying to tell us something with "tree knocks?" Or is there more to it? After reading this book you may wonder.

Just about all the researchers I know believe that knocking on a tree with a wood-on-wood sound will attract a Bigfoot, and some of them claim to have heard a wood knock in return. I have heard it, done it, and recorded the rhythmic sound of tree knocks, and thoroughly believed the sound came from a Bigfoot — never giving it a second thought. I actually got a glimpse of one after some wood-on-wood knocking. It sounded like a signaling, back and forth a messaging to each other. However, I did not actually see the wood knocking and the Bigfoot at the same time…but it was on the same evening.

The close-up knock that I heard in the Sierra Camp in 2011 was almost deafening…more like a 'pop', or a gunshot, within feet of me and during the daytime…one single huge pop. Since then, and since quantum physics has been a more active part of my brain, I take pause and recount the story of a couple different researchers who I think I know fairly well.

Unknown to each other, they both have the same kind of report, e.g., claiming they have personal communications and have a relationship with Bigfoot, and that some of those sounds are not a Bigfoot knocking on a tree. After listening to their stories and thinking there might be another unusual 'something' going on besides just wood-on-wood coming together with a bang, I've given quantum physics a shot.

These people who claim seeing Bigfoots going into, or coming out of a tree are not alone. One gentleman, who I met in a Bigfoot hotspot, told me that he'd seen them in that area on many occasions over the years, but always up in trees. Were they up there merely observing him or could there be another explanation? Do trees play a part in their elusiveness?

On countless occasions, batteries have been known to go dead when these beings are around. For many years I've wondered what role, if any, energy has in their coming and going. And do trees play a part in their elusiveness? A recent discussion with a few of my colleagues made me seriously consider this. If Bigfoots are interdimensional, it may take energy for them to break through dimensions and change their vibrational frequency.

Could they be reenergizing from a tree, where they've absorbed energy for their next interlude with our dimension? Plus, on occasion their appearance happens during a 'lighting strike', suggesting again that energy is involved. Maybe that's why they are rarely seen in the wild, not on a tree, but in a tree — something to ponder.

Glyphs: Natural Stick Structure, or Bigfoot Secret Language?

Some Native American tribes left sticks placed in certain ways. These were intended to represent a readable character in a message to follow a tribal member.

Glyphing by David Claerr

Several researchers I know claim that friendly Bigfoots are leaving them glyphs. It's obvious to many looking at these glyphs they are simply the natural way of the forest; they just fell that way. But some are placed abnormally where there was no sign of human presence. These researchers believe that these glyphs are left for them to interpret.

Unless a human is playing games, these abnormal glyphs cannot be easily explained. They make one ponder — what message did Bigfoot want to leave? Many of these same people also experience 'Mind-Speak'. If this is truly happening, I would call it quantum entanglement. And it also might cause one to wonder; if they can mind-speak, why leave a glyph? One answer is we humans don't often pay attention to what's being relayed to our minds and need a picture; even if it's a few sticks carefully placed in our path.

Portals: Although portals are not an attribute of these beings per say; let's look into it.

It's time now to visit the Woo-Woo camp. By calling it the Woo-Woo camp I don't mean to discredit those folks that ascribe to what most call very strange and absolutely paranormal. That's what this book is mostly about, taking a stab at explaining those unusual accounts. Is there a way that beings who have been seen in this macro-world, can transfer to another space that we don't see — beings who are interdimensional? Are there portals to other dimensions outside of our perception?

For several years I've heard accounts of these creatures disappearing through a portal but didn't want to consider it. However, after delving into quantum physics, I've become aware that there could very well be a recognized science behind such events. Portals are now very seriously studied.

NASA's Goddard Space Flight Center

"A favorite theme of science fiction is "the portal" — an extraordinary opening in space or time that connects travelers to distant realms. A good portal is a shortcut, a guide, a door into the unknown. If only they actually existed. It turns out that they do, sort of, and a NASA-funded researcher at the University of Iowa has figured out how to find them.

"We call them X-points or electron diffusion regions," explains plasma physicist Jack Scudder of the University of Iowa. "They're places where the magnetic field of Earth connects to the magnetic field of the Sun, creating an uninterrupted path leading from our own planet to the sun's atmosphere 93 million miles away."

Observations by NASA's THEMIS spacecraft and Europe's Cluster probes suggest that these magnetic portals open and close dozens of times each day. They are typically located a few tens of thousands of kilometers from Earth where the geomagnetic field meets the onrushing solar wind. Most portals are small and short-lived; others are yawning, vast, and sustained. Tons of energetic particles can flow through the openings, heating Earth's upper atmosphere, sparking geomagnetic storms, and igniting bright polar auroras.

Portals form via the process of magnetic reconnection. Mingling lines of magnetic force from the sun and Earth criss-cross and join to create the openings. "X-points" are where the criss-cross takes place. The sudden joining of magnetic fields can propel jets of charged particles from the X-point, creating an "electron diffusion region." Posted by: Susan Hendrix, NASA, 2015 nasa.gov/mission_pages/sunearth/news/mag-portals.html

Mind Speak:

Telepathy? Does it happen? Over the many years I've been interviewing people I've heard several reports from those who said they were told to do something by a Bigfoot, not in 'words', but by speaking directly to their minds. So I've wondered, did those peoples' imaginations go on a field trip or could it have really happened? To my knowledge, I've never had it happen to me. A mind can do some very imaginative things; it can fabricate and assemble different matters one has previously experienced.

That said, it's not unusual for someone to 'feel' that Aunt Mary was in an accident, the phone rings, and, sure enough, Aunt Mary was in an accident. There are those who are gifted in telepathy, and I'm of the opinion that they are not alone. Is telepathy an ability that we all have? I think so; it just needs to be cultivated. After reading chapter 11, The Pineal Gland, you might agree.

Vocal Ability:

In my first book, Voices in the Wilderness, I mentioned several unusual sounds we at our Sierra Camp, witnessed. One of them specifically, has puzzled me more than the others. It was when we heard our storage barrels, which were unquestionably secured with cables, being ripped from the trees and the supplies in the barrels being tossed around. Later, when we peered out our shelters' opening and looked, nothing had changed. This left us trying to figure out what we thought we had heard. Our choices were: (1) We all, collectively, just thought we'd heard those sounds, or (2) We did hear those sounds.

After contemplating this, we thought of a couple other choices: (3) These creatures are capable of mass hypnosis, (4) we were listening into another, perhaps parallel, dimension, or (5) back to number 2 — they created the sounds with their complex vocal mechanism. From mimicking names, a car door slamming eight miles deep in the wilderness, to hearing a herd of horses stampeding toward camp, all these enigmas may have something in common.

Bigfoot's vocal mechanism isn't just complex, it seems extremely complex. If this is correct, my question is, "could they actually make a sound that would vibrate at a frequency above the speed of light?" So with this, my "sound and light" research began.

Sound to Light

"Visible light is an electromagnetic wave, or particle, consisting of oscillating electric and magnetic fields traveling through space. The frequency is the number of waves that pass a point in space during any time interval, usually one second. We measure it in units of cycles (waves) per second, or hertz. The frequency of visible light is referred to as color, and ranges from 430 trillion hertz, seen as red, to 750 trillion hertz, seen as violet.

Researchers at the Lawrence Livermore National Laboratory in California successfully converted sound waves to light radiation by reversing a process that transforms electricity to sound, which is commonly used in cell phones. This is the first time that sound has been converted to light. The findings were published in Nature Physics, 2009, and could improve how computer chips, LEDs, and transistors are made, and also have applications in ultrafast materials science and terahertz radiation (T-ray) generation."

Source: popsci.com/scitech/article/2009-03/sound-becomes-light.

For years I have wondered what these beings are doing with and around electricity. At our Sierra Camp, new batteries would often go dead. The last time this occurred was in 2011. I was at camp alone and had just put fresh lithium batteries in the recorder, yet when I hit the button, I got nothing, dead! This also has happened to Scott Nelson (Crypto-Linguist) on occasions when we were in camp together.

Al Berry mentioned his batteries going dead at our camp, but blamed it on the cold (Volume 1 of the *Bigfoot Recordings*). I've noticed on several sighting reports that Bigfoot was seen along power lines and near, or on, power transformers. I believe something is happening with energy-power and these beings. An "Out of the Box" hypothesis; their vocal ability can produce ultrasound but they need electricity. That high sound frequency can change into light and that light frequency can go out of our visual perception.

Everything you can see, hear, or touch seems real when there is light and its frequency is lowered (or raised) into our perception of reality, i.e., the lights' frequency. When we leave our bodies our frequency (Energy) is raised to a level of non-perception to the human eye. At that time we are energy only and on our way to something else. Hallelujah.

Can Bigfoots Disappear?

Over the years I've received many reports claiming that — Wait, I've got a whole chapter on this (Chapter 16, Invisibility), but I suggest reading the chapters in chronological order to fully appreciate that aspect of these most unusual beings.

Chapter 4
Quantum Physics & Spirituality

What is the God Particle?

CERN, the European Organization for Nuclear Research, near Geneva, Switzerland, houses the Hadron Collider. It's the world's largest and most complex scientific instrument used to study fundamental particles, the basic constituents of matter.

The Higgs Boson Particle was discovered in July of 2012, and it holds exciting information. Physicists believe the Higgs Boson particle is the material of the universe prior to the Big Bang and helps give other particles their mass. In fact some believe it is responsible for the Big Bang. Therefore, it was named the God particle. This discovery could lead physicists to better understand what the universe was prior to God's saying, "Let there be Light."

CERN Large Hadron Collider

"First of all, the Big Bang wasn't very big. Second of all, there was no bang. Third, Big Bang Theory doesn't tell you what banged, when it banged, how it banged. It just said it did bang. So the Big Bang theory in some sense is a total misnomer."

Dr. Michio Kaku, American (b: 1947), theoretical physicist, Harvard, UC Berkeley.

Author's note: "If nobody was there to hear it bang — did it really bang?" Actually, I think the Big Bang happened when God decided to create Dark Matter, which I believe contains all knowledge, a variety of light photons, and started a limited dimension of observability.

We only observe in three dimensions; all due to the fourth dimension—light. Light has its own vibrational frequency and is limited in scope — simply, to date, we can't see beyond light frequency parameters (see Chapter 7, Vibrations).

Following quote from the University of Oregon: abyss.uoregon.edu/~js/21st_century_scinece/lectures/lec03.html:

"Plato presented the foundation of natural philosophy (science) with his mathematical ideas. Later Aristotle pioneered the implementation of science to everyday events. The core of Aristotle's philosophy is that nature is understandable, explainable and predictable. However, he developed a holistic view of nature, versus the Newtonian mechanical view of science. In Aristotle's philosophy objects are assigned destiny and a teleology concept."

Sir Isaac Newton (1643-1727) and Albert Einstein (1879-1955) are regarded by many as the forefathers of modern physics, both held theories that are fundamentally different from the other. So in the grand scheme of things, who was more correct? Einstein or Newton?

In the world of Newtonian physics, everything looks the same to everyone else in the universe, irrespective of your location and speed. I don't know about the rest of you, but this seems like a very logical concept, probably because this is how we all view everyday life.

Newton lived in a time of hand written calculus. This seemed to limit his view of space-time as flat, unchanging and very boring, but that is not at all the case in Einstein's world. To Einstein, space-time is very dynamic, changing depending on gravity and velocity. (Source) Futurism.com

1686. Newton founded his principles of natural philosophy on three proposed laws of motion: the law of inertia, his second law of acceleration and the law of action and reaction; and hence laid the foundations for classical mechanics. Wikipedia

1900. The birth of Quantum Theory: German physicist Max Planck published his groundbreaking study of the effect of radiation on a "blackbody" substance, and the quantum theory of modern physics is born. In 1918, Planck was awarded the Nobel Prize in physics.

Einstein, however, is best known in popular culture for his mass-energy equivalence formula $E=mc^2$ which has been dubbed "the world's most famous equation".

For his work, Einstein received the 1921 Nobel Prize in Physics." End of quote.

So how does quantum physics relate to spirituality and Bigfoot? In my 45 plus years of researching the Bigfoot phenomenon I have heard several very strange reports. Most of these reports, from very earnest people, claiming that these creatures can, and do, disappear — actually, vanish from their sight.

Is this even possible? Can the laws of quantum physics give us the answer? They work throughout the universe, on this planet, within each of us, and also in Bigfoot. The accepted mathematics of quantum physics says that there is much more going on than meets our three-dimensional eyes.

Many people have claimed that they saw a UFO take off in a flash, had unusual lights follow them at night, heard unrecognizable sounds when nobody was around, and the list goes on. In an effort to understand these reports, together with what took place at our Sierra Camp, I began my research with the principles of ordinary science, which seemed ordinarily reasonable at the time.

A Bigfoot family was interacting with us. It was real, definitely exciting, sometimes frightening, and at the same time very puzzling. There had to be a normal explanation for what we were experiencing. We just had to find it. Considering their huge footprints, along with the distance between them, and those enormous sounds they made, these giants had to be massive.

How could they move so fast — in and out of our sight in a flash, leaving only tracks and the vocalizations we captured as evidence? How did they stay so hidden? Although we tried several times we couldn't capture a picture, they just weren't having it. Surely they didn't understand what a camera does; or did they?

One man in the group was frightened off — it was just too strange and certainly didn't fall into the beliefs of his religious background. Hybrid monsters and biblical giants do not really exist in this day and age; or do they? I believe he must have thought that they must be some kind of demon to do what they do.

Al Berry at Sierra Camp, 1972

Al Berry, an investigative reporter who came to the camp to uncover what he thought was surely a hoax, had a Master's degree in science. I think he was always hoping to uncover a hoax. Al thought our reports simply had to be one. He often told me to not talk about the unusual happenings that we were experiencing; he said that it would disrupt our credibility and the scientific community would give the whole story a big boot right out the door. So for years, very much like the others in the group, I only talked about those strange happenings to my family and close friends.

Many of these accounts are mentioned in my first book, *Voices in the Wilderness*, which I first published in 2012 and the second edition in 2013. Suffice it to say that after more than four decades the answers to many of those questions still remain stealthily concealed under the Sierra Mountains' limbo log. But in the last few years, I've gotten more limbered up, bought a rake, and decided to get under that log.

I believe that any question and/or enigmas can, by some scientific method, be answered. So where does anyone turn when we can't seem to find an answer? We turn to someone who we think might know the answer and listen.

> "Our greatest weakness lies in giving up. The most certain way to succeed is always to try just one more time. I have not failed. I've just found 10,000 ways that won't work."
> Thomas A. Edison (1847 – 1931) American scientist and inventor.

> "If quantum mechanics hasn't profoundly shocked you, you haven't understood it yet. Everything we call real is made of things that cannot be regarded as real."
> Niels Bohr (1885 – 1962) Danish; received Nobel Prize in Physics 1922.

> Einstein said, "Science without religion is lame, religion without science is blind."

I really liked that. Not to put words into his mouth, but I think if Einstein were to say it today, he would use the term spirituality and not religion. Religion, in my opinion, will often inadvertently keep people from a true connection with their spirituality. Of course, there are exceptions to this.

These religious people mean well and for the most part are loving, caring individuals, but can easily get caught up in an indoctrinated belief that is partly, and I think debatably, manmade. Many were brainwashed to think that, due to their horrible sins at the age of four, they had to be saved from the pits of hell.

As their paradigm developed many got stuck on a biblical scripture that warns if anyone adds a single word to these teachings they will be punished by God (Galatians 1:9). Think about it. "God is Love." (1 John 4:8) Right? How many different interpretations of the Good Book have we had since that chapter in Galatians was canonized?

For hundreds of years, the books of the Bible have passed through the hands of man and was altered a bit to suit kings of the time. However, I still turned to it after first encountering these beings in the 70's. I didn't know of another book that would have more information about giants? I had to wonder, could the type of giants that existed then still exist now?

Science's Microscope

For years I've been speaking at conferences and symposiums also attended by well-meaning scientists, some who I believed to be spiritual; and, also wanted to believe that the Bigfoot creatures exist. But, these same scientists were true to their profession, they wanted, and still want, hard evidence such as bones or a body; something they can put under a microscope, dissect on a laboratory table, and in general, "thoroughly study it."

So until recently, learned scientists would not give my Quantum Bigfoot theory a second thought. Hard evidence is difficult to come by when it comes to Bigfoot and after decades of looking and hoping for those bones, or a body, those few and far-between scientists seem to be opening their minds to a different type of science, the science that takes over when classical science seems stalled.

After all, science is supposed to be self-correcting. Classical science requires a body of evidence, it can't be predicated on likes or dislikes, or be founded on a belief. All assumptions must be supported; and more important, must be proven by repeatable experiments.

Although I feel I know some of these scientists personally and call them my friends, they didn't want to be on the outside of their disciplinary training with a subject such as Bigfoot. Academia is a tough business, who wants to stick their head out for an institution or discipline to chop it off?

A scientist can believe in a supreme being (God, whom they've never seen), but still, have a strict classical discipline when it comes to their work. The cartoonish sounding subject of Bigfoot can easily affect their funding when it comes to other projects they want to pursue. They simply won't get money; the old dictum of "publish or perish" is still the play-by-the-rules in science. In fact, they could lose tenure putting their retirement in jeopardy.

Quantum Physics & Spirituality

Dr. Grover Krantz

The following quote is from Dr. Grover Krantz (1931-2002) professor of Anthropology, Washington State University, Pullman:

> "It is not surprising that many contributors failed to make clear, affirmative indications of their opinions as to the reality of the Sasquatch. If anything, it is surprising that so many did come out in favor of its existence. To assert such an opinion can be dangerous to one's academic career: I can state this from personal experience. There is a justified fear that employment might not be continued, raises and promotions may be denied or delayed, and working conditions otherwise be affected in an adverse manner."

That said, the Bigfoot phenomenon seems to have captured the curiosity of certain professors and won't let go. They read and hear of just too many sincere people chiming in with credible sounding, albeit sometimes very unusual, reports. But Bigfoot seems to stay one step ahead of being exposed. I don't know of any research method, e.g., tracking, trail cams, drones, etc., being utilized today that is getting the results professionals think are desperately needed. So, how do we find the answers?

Einstein, along with other physicists, continued the study of quantum physics, and its benefits go undisputed. We now use it every day in the micro-world, e.g., microwave ovens, cell phones, televisions, MRI Scans, etc. But why don't we consider that its laws also exist in the macro-world that we live in? Or, maybe we just need to discover how that could be done. Could Bigfoot have discovered it, or somehow just know about it?

Quantum Physics & Spirituality

Sir Roger Penrose
OM FRS
English mathematical physicist (b. 1931)

"Sir Roger Penrose, Professor of Mathematics at Oxford University, believes he has identified the secret that keeps the quantum genie tightly bottled up in the atomic world, a secret that was right in front of us all along, i.e., gravity. In his novel view, the same force that keeps us pinned to the ground also keeps us locked in a reality in which everything is tidy, unitary, and—for better and for worse—rooted in one place only. Everything we see with our eyes is considered 'Local'. The world of quantum mechanics is considered 'Non-local' because we can't see what's going on with our eyes."

From: Discover: timfolger.net/penrose.pdf

From chemwiki.ucdavis.edu: "Classical mechanics accurately describes most systems that can be easily observed and measured. Objects that are a "normal" size (larger than a molecule and smaller than a planet), at a "normal" temperature (anywhere close to room temperature), going a normal speed (anything significantly less than the speed of light) fit the models set forth in classical mechanics. It is only when the system being observed begins to violate these parameters that quantum factors come into play. An important aspect of the quantum mechanical models is the fact that as the conditions approach "normal" the quantum mechanical model approaches the classical model."

"An important scientific innovation rarely makes its way by gradually winning over and converting its opponents. What does happen is that its opponents gradually die out and that the growing generation is familiarized with the ideas from the beginning."

Dr. Max Planck

Dr. Max Karl Ernst Ludwig Planck: (1858 – 1947) Father of modern quantum theory; 1918 Nobel Prize Winner in Physics.

Scientists are now discovering, through physics, that empty space (Dark Matter, Dark Energy) is not actually empty. However, they do exist in a dimension outside the human light sensing spectrum; and, our observable vibrational frequency (Chapter 7, Dark Matter). It seems to me that classical science has restricted itself by its own disciplines and because of those disciplines, may never grasp the big picture. If we use the classical box to try and determine all that exists, we couldn't even begin to understand the cosmos and the hidden world of spirituality.

The math of quantum physics indicates that there are at least eleven dimensions in existence, and possibly innumerable dimensions. So, could the laws of quantum physics be the answer to the Bigfoot mysteries? If so, how do we move forward in that possibility?

Albert Einstein is still the most recognized physicist in the world. His genius, although questioned by some, is much of my inspiration for this book. It appears to me that quantum physics and spirituality are synonymous. So I say, "If Einstein thought science and religion should work together, I agree; if it was good enough for Einstein it's good enough for me." Let's find out how these giant beings, we refer to as Bigfoot, do what they do!

Quantum Physics & Spirituality

1. "I believe there are no questions that science can't answer about a physical universe."
2. "I hope I have helped to raise the profile of science and to show that physics is not a mystery but can be understood by ordinary people."
3. "Look up at the stars and not down at your feet. Try to make sense of what you see, and wonder about what makes the universe exist. Be curious."
4. "Intelligence is the ability to adapt to change."

Dr. Stephen Hawking

Stephen Hawking (b: 1942) CH, CBE, FRS, FRSA; English theoretical physicist and author.

Chapter 5
Ancient Texts and Quantum Physics

The Vedas are a large body of knowledge texts originating in the ancient Indian subcontinent. Composed in Vedic Sanskrit, the texts constitute the oldest layer of Sanskrit literature and the oldest scriptures of Hinduism. Hindus consider the Vedas to be "apauruseya," which means "not of a man, superhuman" and "impersonal, authorless." SlideShare/Hindu-books: www.lonweb.org/links/sanskrit/lang/007.htm

Schrödinger, Einstein and Tesla were all Vedantists. Bohr, Heisenberg and Schrödinger regularly read Vedic texts (source: Sacred texts). These physicists were required to search outside the parameters of classical science for answers to how all things work.

> "Religious traditions are often criticized for providing untestable elements in their books of wisdom. Among these elements, the soul's existence is one of the most debated in talks between science and religion. But is the concept of the soul really so vague and does it really have no empirical evidence for a practical theory? Vedic philosophy and quantum mechanics may bring interesting ideas to consider.
>
> Quantum chromodynamics is the theory describing sub-nuclear particles called quarks and gluons, which are the constituents of hadrons such as protons and neutrons. It is part of the Standard Model describing our current understanding of elementary particles. One interesting feature about this theory is that even though it was postulated in 1964 by Gell-Mann and Zweig, and is widely

accepted by the physics community, to this day all experiments validating the existence of the quarks have produced only indirect proof of them.

In fact, according to the theory itself, no one really ever expects to see direct proof of them at all, viz. free quarks. Pulling two quarks apart requires so much energy that you create new particles in the process which attach themselves to the old quarks and thus you never end up with free quarks to detect. The important point here is that the experiments that confirmed the existence of these quarks never actually detected them directly and no one really demanded such a proof.

They saw certain particles coming in and certain particles coming out according to what the theory predicted, but never actually saw the free quarks by themselves. Therefore, indirect detection with no expectation of direct detection is still valid to confirm the existence of something in science.

On the other hand, Vedic philosophy tells us about the existence of another type of particle called the atma, the soul, the self, having different properties from those of ordinary particles. This is not unusual in physics, as is the case with antimatter particles.

To declare that there is no soul, just because we can't directly see it with our current instruments, is akin to saying that there are no quarks because we can't detect them directly in our labs. Both the soul and the quarks' existence can be inferred indirectly by their effects (living symptoms and consciousness in the case of the soul; particle interactions in the case of the quarks).

Ancient Texts and Quantum Physics

Even though there are other theories in science describing the origin of consciousness and living symptoms, these remain very unsatisfactory and many times philosophically dismissive. Furthermore, the wealth of data regarding topics like reincarnation gathered by careful scientific studies such as those of Dr. Ian Stevenson (1918 – 2007) at the University of Virginia School of Medicine can hardly be accounted for by any explanation other than one involving some notion of what is here referred to as atma." Source: the huffingtonpost.com/ vedic-philosophy-and-quantum-mechanics

"Some physicists proclaimed that quantum physics is a boundary science, a science that points to the boundary which humans can reach while researching the physical universe. In other words, these physicists assert there is something beyond that boundary that science will never be able to identify." Source: Dr. Jeffrey Satinover (b: 1947) American, MD & PhD in physics. Quote from the International Kabbalah Congress, Israel.

Energy cannot be created or destroyed. It can only change from one form to another. For me this is simultaneously quantum physics and spirituality — quantum physics merely being the laws by which things are governed. Jesus turned water into wine, healed the sick, raised the dead, and told His apostles they could do those same things. Although Jesus was worshiped, he never said, "Worship me", he said, "Follow me." This is important to think about...it's quoted eleven times in the four gospels of the New Testament. God knows how important His Word (Christ) was to humanity, and what He said should be our focus.

Ancient Texts and Quantum Physics

Some have managed to wrap their heads around the concept of quantum physics, but if you are one of the people who may not have even tried, or been exposed to it, you may not understand how I think it could relate to the giants who stay in the woods and remain hidden.

The world of quantum physics also provides an explanation for those who believe in God; how this divine energy infiltrates the world we live in. If we allow our minds to really capture this thought we all win. Many religious people have never before considered quantum physics as the way God works throughout the universe, on this planet, in you and in me.

Often I'm asked if I'll be a guest on a paranormal talk show. I'll usually say yes, but I really don't like to use the word 'paranormal.' In a way it makes me feel weird. So that was another reason I turned to quantum physics; the science that I believe is behind many mysteries of today, including the Bigfoot phenomenon and unusual things I've witnessed and heard of.

When we get into definitions, quantum physics says, "Nothing is real until it is observed." That was difficult for me to get my head around. However, a scripture in the Good Book says, "Faith is the substance of things hoped for, the evidence of things not seen" (Hebrews 11:1). To me, these two axioms indicate the same thing, just different words from different times. Is 'Faith' actually a substance waiting to be observed to make it real?

We do not see with our eyes how all things work, the idea must be conceived inside our heads first. It is a bit like trying to conceive the end of outer space. Where could it possibly end and how could it end? What would be at the end? A no-end, continuing cosmos, cannot be understood using our cardinal thinking. We can only conceive it and understand that it has to be that way. In order to conceive many things, we must connect with something inside, i.e., be 'Born Again' (John 3:3).

In the very center of our brain is a gland called the Pineal. It's considered by the ancient Hindus as the seat of the soul, depicted in Egyptian hieroglyphics as the Eye of Horus, and has many of the same attributes as the physical eyeball; the more common term being, "The Third Eye." The big difference is that it doesn't require outside light.

The Pineal gland has been credited for giving us our perception, instincts, and awareness of things that we don't often see with our three-dimensional eyes. I've noticed that babies, prior to being indoctrinated into civilization, seem to have awareness, but that innocent awareness begins to leave them after a couple years of being around us know-it-all adults.

I believe that a babies' pineal gland is more active during those early months; they seem to have a "knowing" in their eyes, but a baby's sapience hasn't yet developed. Wouldn't it be good if they could talk and tell us what they know about where they came from? (Matthew 18:3, Ephesians 1:4).

In most humans, unlike animals, our pineal gland gets calcified by the processed foods we eat, a toxic environment, and the controlled water provided by most public services, fluoride being the big culprit. By giving ourselves the proper organic diet, among many other things, we can begin to decalcify that gland and maybe become a people that can better listen to their instincts.

Rarely do we follow those intuitions; we analyze using perception sent to us from what our senses, our eyes, ears, nose, and feel detect. Some people believe the pineal gland is the catalyst that allows us to communicate with everything, including God (the

Ancient Texts and Quantum Physics

supreme alien) —who said we were made in His image. (Find a more thorough scientific article on the Pineal Gland in Chapter 11).

Is there really a physical receptor that allows us a connection to another realm of reality; something not really in woo-woo land, or that can only be found in church on Sunday by bowing our heads or looking up at the ceiling? What is it we are looking for, something outside of ourselves to come down, swoop into our soul and heal our heartbreak, or answer difficult situations like financial problems? All of the power and knowledge we need to make anything change has been given to us.

> Jesus said, "...whatsoever thou shalt bind on earth shall be bound in heaven; and whatsoever thou shalt loose on earth shall be loosed in heaven." (Matthew 16:19, John 14:17).

Are we listening to that voice within us that speaks to our subconscious? The Bible says that we have guardian angels who watch over us, often trying desperately to communicate something to our inner-being (via the pineal?). On countless occasions, I have no doubt that something, or someone unseen, changed a circumstance that I was in. I can't count the times I've been in serious trouble and suddenly the condition changed. Living on the edge, as I've done so much, the common assumption is that I should have died many times — just wasn't my time to push up daisies. Maybe I was supposed to write a book or something!

But for whatever reason, I am here writing about aliens, angels, Bigfoot, and quantum physics. In my opinion, there is no difference between angels and aliens, some are good and some are definitely bad. Rarely have I found anyone who thinks we are the only beings in the universe. Unidentified objects in the sky are commonplace in Central and South America. Actually, they've been reported all over the world and are still played down by governments. As more and more reports come in, there seems to be an awareness (enlightenment) seeping into our consciousness.

We are not alone in this vast universe. Using our known methods of travel, how could anything, or anyone, possibly get anywhere far into our galaxy within a lifetime, even if they could travel at the speed of light?

Depending on the curvature of the earth, on the average, it takes eight minutes and twenty seconds for light from the sun to strike the earth. If we see an explosive flare on the sun it happened over eight minutes before! When astronomers dictate speed in the Cosmo, they refer to it as light speed; that is how far an object is in space from earth traveling at the speed of light. In order to get to any place of great distance, beings who want to go far-far-away must travel faster than light.

Einstein, along with other physicists, agrees that mass cannot go faster than the speed of light (186,300 miles per second). I believe Einstein meant nothing we can see in our three-dimensional world can exceed the speed of light. Once an object goes faster than light it can no longer be observed by our eyes. The object turns into pure energy.

In 2014, physicist James Franson of the University of Maryland threw a curve into $E = mc^2$ when he theorized that Einstein was wrong. Our thoughts travel faster than light, but thoughts are not mass — just energy.

Can our thoughts actually make anything happen? Scientifically speaking, "Yes," (see Chapter 9, Quantum Entanglement), biblically speaking, "Yes," Matthew 12: 35-37."

"It is through science that we prove, but through intuition that we discover." Poincaré

Jules Henri Poincaré (1854-1912) was a French mathematician and theoretical physicist, engineer and philosopher of science.

Jules Henri Poincaré

With the above statement from Poincaré, who Einstein looked to, the author states the following.

"Mass cannot exceed the speed of light, but it can actually continue as its true self, in the form of energy. When energy in the form of an object meets and then exceeds the speed of light it ceases to exist as something observable. Mass is energy; only visible when its speed is within lights' velocity."

So, when Einstein wrote, "What we have called matter is energy, who's vibration has been so lowered as to be perceptible to the senses," I can get my head around those words.

Mass is energy; only visible when its speed is within lights' velocity. Now I understand, or more precisely, I think, speed beyond light velocity is how aliens move, travel immense distances, rapidly appear in and out of our sight, and maybe, just maybe, is the "how" for Bigfoot to appear and disappear. Through brilliant reason, Einstein called 'light' the fourth dimension (Chapter 13, Quantum Time). The Bible says that God is the light of this world (John 8:12).

God said, "Let there be Light." When the Big Bang happened, there was nothing but Dark Matter. All matter is made of particles. At their smallest level, they are an actual wave. Learned scholars label this phenomenon of physics "String Theory." Those waves vibrate all the time. They are the constant, never-ending energy that everything is made of. And, as mentioned in the previous chapter, the Higgs Boson particle, the God particle has no mass and was here prior to the Big Bang.

"All matter arises and persists only due to a force that causes the atomic particles to vibrate, holding them together in the tiniest solar system, the atom. Yet in the whole of the universe, there is no force that is either intelligent or external, and we must, therefore, assume that behind this force there is a conscious, intelligent Mind or Spirit. This is the very origin of all matter."

Planck as cited in Eggenstein 1984, Part 1: See "Materialistic Science on the Wrong Track."

It's been said that the universe is expanding; therefore our consciousness should be expanding too. Jesus said He was going to prepare a place for us. (John 14:2) That is exciting to me. But it's been 2000 years, how much time does He need? Maybe He's thinking, How much time do humans need?

Humans get very complacent, going day-to-day overlooking the fact that at any time a catastrophe could take place. And, eventually, another planetary catastrophe of some sort will. Do you have your handy-dandy survival knife sharpened and shelves of food stored? Many people do. Someday our mundane day-to-day existence will change dramatically. The change will be good for some and very bad for others. So who is actually getting ready?

We live in a passive cultural environment, created for us by us. Most think it's good to have tangible things for comfort, but is it really? Obviously, animals don't care about those things and Bigfoot probably doesn't either. When a major disaster hits, the animals, including Bigfoot, will know what to do. Will they help us, or will they simply take over?

Chapter 6
Vibrations

"If you want to know the secrets of the universe, think in terms of energy, frequency, and vibrations."

Nikola Tesla

Nikola Tesla: (1856 – 1943) Serbian – American engineer, physicist, and futurist.

The science of vibrational frequency

From Dr. Planck: "All the physical matters are composed of vibration. As a man who has devoted his whole life to the most clear-headed science, to the study of matter, I can tell you as a result of my research about atoms this much: There is no matter as such. All matter originates and exists only by virtue of a force, which brings the particle of an atom to vibration and holds this most minute solar system of the atom together. We must assume behind this force the existence of a conscious and intelligent mind. This mind is the matrix of all matter."

Vibrations

It's not paranormal, it's not supernatural, and it's definitely not unscientific...it's just quantum physics at work. Too many people want to 'brand' the unusual reports of Bigfoot and put them into Woo-Woo land. Prior to doing that, I suggest delving into quantum physics, and please don't tell me that it's all just 'theoretical', e.g., Quantum Theory.

Because the existence of quantum mechanics is studied in the micro-world, much of quantum theory is established through mathematics and you won't ever see the physical properties, you can, however, receive its benefits. Classical science establishes theories through observing duel experiments that prove each other. They do that within the confines of our limited four dimensions (3+1, the 'one' being light, or 'c', in Einstein's $E=mc^2$). Since most of the quantum theory is established outside our limited dimensions, one can never prove a theory in physics; physicists can only validate it or falsify it through mathematics.

> From Ari Lakakis, *Earth Unchanged*: "In 1905 Albert Einstein made a discovery that ignited humanity's curiosity. Einstein proved through quantum physics that energy and matter are interchangeable and interconnected. The universe and all things within it are interchangeable. Your energy, vibrational frequency, and physical self are interchangeable. We are one universal energy body manifesting itself in various forms throughout the cosmos.
>
> Matter can be broken down into smaller components. These smaller components can transport us beyond the physical realm and into the realm where everything is pure energy.

Vibrations

When two frequencies come together, the lower one rises to meet the higher. We call this the principle of resonance. Example: When a piano is tuned, a tuning fork is struck and brought close to the piano string that carries the identical musical tone. The string then raises its vibration automatically and adjusts itself to the same rate at which the fork is vibrating."

These days, scientists recognize that countless vibrations radiating at various frequencies control the molecules in your body. Now, although most frequencies exist outside of your normal range of perception, you generally observe them as both color and, or sound. Seven colors exist within a rainbow and seven notes within the color and a musical scale. For example, you can hear the color blue as the musical key of D, which vibrates at 587 Hz. Here's the fascinating part. If a frequency vibrates fast enough it emits as a color of light. Therefore, if you want to convert sound to light, you would simply raise its frequency forty octaves. This results in a vibration in trillions of cycles per second.

Thus, if a pianist could press a key far above the eighty-eight keys that exist on a piano, that key would produce light. They would produce a chord of light in the same way they create a chord of sound. The only difference is you would see those chords as colors of light because they would be traveling at the speed of light.

When you use the principle of resonance, you can accelerate the speed at which your molecules vibrate. When the atoms within you lose speed, you create third-dimensional matter. However, when they accelerate, your consciousness reaches a higher dimension. Hence, the higher you raise your vibrational frequency, the faster you expand your consciousness and the closer to spiritual enlightenment you become. We call this the universal Law of Vibration." End of quote.

Everything is energy. The Law of Vibration states that in the universe energy is continuously moving, vibrating, and shifting into various forms— we, including Bigfoot, are no exception. These vibrations vary. If they're low, they're slow. If they're high they're vibrating fast. Just think; nothing ever stands still and nothing ever truly rests.

> "Before the invention of specialized devices, people would have labeled you clinically insane if you told them that a force exists between two magnetic bodies making them attract or repel one another. Why? Because, they could not see the interaction with their eyes. Today, we accept this information as fact. We don't question it. I think the day will come when humanity as a whole accepts that our thoughts, actions, and surroundings emit a vibrational frequency that can both hurt or help."
> Source: *The Hidden Lighthouse*, Word Press

In 1910, Wallace D. Wattles wrote a book called *The Science of Getting Rich*. He speaks of a way to 'think,' which can form all things:

> "Everything you see on earth is made from one original substance, out of which all things proceed. It is a thinking stuff from which all things are made, and which, in its original state, permeates, penetrates, and fills the interspaces of the universe. A thought, in this substance,

produces the thing that is imaged by the thought. Man can form things in his thought, and, by impressing his thought upon formless substance, can cause the thing he thinks about to be created."

Hmm, is this how Jesus did His miracles? (Hebrews 11:1). An intricate universal web of energy connects everyone on earth through various frequencies of vibration. In addition to that, every one of us has a distinctive aura that vibrates at its own unique vibrational frequency. Simply put, humans and animals alike radiate the same great energy that created the cosmos. At the most fundamental level, we interconnect intricately and divinely.

When you alter your vibrational frequency, you directly affect your physical world. Hence, when you're having a bad day and everything goes wrong you think, what is happening today? Is the whole world looney tunes? The answer is no, it's not; it's your vibrational frequency in a horrible state.

Now, as you know your vibrational frequency affects your reality. The various mental, physical, emotional, and spiritual states that you experience are simply different levels of vibrating energy and frequency. As a result, thoughts, and feelings of fear, hatred, envy, anger, grief, and despair vibrate at a slow and low frequency, while thoughts and feelings of love, happiness, joy, and gratitude vibrate faster and higher.

Could this be what the Bible is talking about in 1 Corinthians 13:13: "And now abides faith, hope, love, these three: but the greatest of these is love."

This is exactly why Jesus taught about Love and caring for others. Love is the 'key' to everything good.

Vibrations

From Lakakis', *EarthUnchained*: "If you fill your life with negativity, eat rubbish food, have too much alcohol, or even legal drugs, then your energetic field turns cloudy, murky, and jammed. Your vibrational frequency is the magnet that attracts your experiences. And that's why you're responsible for how you choose to exist on this earth, what you choose to experience, and how you choose to evolve and grow in your life. Your free-will to choose in life will never be taken from you. So relax, take a deep breath, and listen to the inter-person.

Diaphragmatic breathing allows you to find a calm vibration, which raises your vibrational frequency as a whole. If you learn how to do it, it can have a direct effect on your entire nervous system, sending you into a constantly calm and tranquil state. It's a skill, something you learn with practice. Ultimately, you want to make a habit of breathing large amounts of oxygen through your diaphragm and into your vital organs, instead of breathing shallow breaths from your upper chest. It takes a bit of work, but it's worth the rise in vibration energy.

Laughter is an energetic activity (and extremely contagious.) You don't need to laugh at anything in particular but simply laugh for the sake of laughing. Another added bonus of laughter—it can lower pain and stress levels. The best part is it's free."

So can Bigfoot disappear? No, it just seems that way to our eyes. They, some of them anyway, can raise their vibration to a frequency that is outside of our perception. (More about this in Chapter 15, Invisibility).

Vibrations

Hopefully, we're getting closer to understanding the laws of quantum physics. Maybe at least now we won't be so quick to call folks delusional when they report strange occurrences of cloaking around these beings. At the very least it should give us pause.

"The day science begins to study non-physical phenomena, it will make more progress in one decade that in all the previous centurions of its existence."
Nikola Tesla

The author on his horse, "JR" and pack mule "Rabbit" descending from Sierra Camp.

Chapter 7
Dark Matter

"We can't see it, but it's there."
The author circa 2010

Understanding particles that exist outside of our normal perception is, in my opinion, important. Do they relate to Bigfoot? If these beings can, at will, change their vibration to a frequency that is outside of our perception I say 'yes' it might relate. In this chapter I hope to further your understanding, or inspire the realization that there's a lot that exists that we cannot observe with our two eyes.

A black hole is one of them. Photographs do not portray black holes. We "see" black holes from astronomical observations, in many spectrums, of a black hole's effect on cosmic materials and energies around it. But for our discussion, a black hole is black and appears empty, but indeed it is not empty.

Three theoretical physicists, Dr. Kip Thorne, Dr. Steven Hawkins, and Dr. John Preskill made a bet regarding information loss in a black hole. (Hey, when you're a theoretical physicist, you have to bet on something besides the Super Bowl.)

Anyway, Thorne and Hawkins theorized that nothing could escape a black hole. However, Preskill of Computer Information Technology fame, was right and he won the bet as reported in "Physics World," July 22, 2004. In a black hole matter and energy are not lost; they are preserved and sent back into space as radiation energy (Dark Matter).

You may be wondering, "Do I really need to know physics to know more about Bigfoot?" Well, we're looking for answers to things we are not seeing, but actually exist, so, "Yes," I think so. Physicists are just now trying to understand Dark Matter and Dark

Energy. After reading this chapter you may feel yourself getting bogged down with this stuff. If that happens, like a photon, zip past the text; but at least consider my hypothesis at the end.

"The formulation of a problem is often more essential than its solution, which may be merely a matter of mathematical or experimental skill. To raise new questions, new possibilities, to regard old problems from a new angle, requires creative imagination and marks real advance in science." Albert Einstein

Quantum physics sprang in part from little puzzlements about how heat is radiated. So, how much may be learned by resolving today's much deeper confusions about dark matter and dark energy?

As the physicist, Niels Bohr said, 'No paradox, no progress."

Source: ngm.nationalgeographic.com

Niels Henrik David Bohr
Nobel Prize in Physics 1922

Dark Matter

From: **NASA SCIENCE: DARK ENERGY, DARK MATTER**

"In the early 1990s, one thing was fairly certain about the expansion of the Universe. It might have enough energy density to stop its expansion and recollapse, it might have so little energy density that it would never stop expanding, but gravity was certain to slow the expansion as time went on. Granted, the slowing had not been observed, but, theoretically, the Universe had to slow. The Universe is full of matter and the attractive force of gravity pulls all matter together. Then came 1998 and Hubble Space Telescope (HST) observations of very distant supernovae that showed that, a long time ago, the Universe was actually expanding more slowly than it is today. So the expansion of the Universe has not been slowing due to gravity, as everyone thought, it has been accelerating. No one expected this; no one knew how to explain it. But something was causing it.

Eventually, theorists came up with three sorts of explanations. Maybe it was a result of a long-discarded version of Einstein's theory of gravity, one that contained what was called a "cosmological constant." Maybe there was some strange kind of energy-fluid that filled space. Maybe there is something wrong with Einstein's theory of gravity and a new theory could include some kind of field that creates this cosmic acceleration. Theorists still don't know what the correct explanation is, but they have given the solution a name. It is called dark energy.

WHAT IS DARK ENERGY?

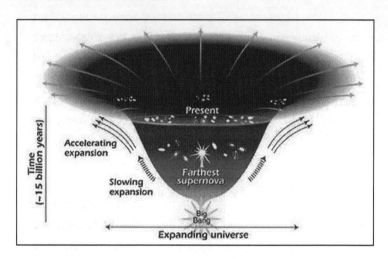

Universe Dark Energy - Expanding Universe

This diagram reveals changes in the rate of expansion since the universe's birth 15 billion years ago. The more shallow the curve, the faster the rate of expansion. The curve changes noticeably about 7.5 billion years ago when objects in the universe began flying apart at a faster rate. Astronomers theorize that the faster expansion rate is due to a mysterious, dark force that is pulling galaxies apart."

Continuing with NASA/STSci/Ann Field: "More is unknown than is known. We know how much dark energy there is because we know how it affects the Universe's expansion. Other than that, it is a complete mystery. But it is an important mystery. It turns out that roughly 68% of the Universe is dark energy.

Dark Matter

Dark matter makes up about 27%. The rest - everything on Earth, everything ever observed with all of our instruments, all normal matter — adds up to less than 5% of the Universe. Come to think of it, maybe it shouldn't be called "normal" matter at all since it is such a small fraction of the Universe.

One explanation for dark energy is that it is a property of space. Albert Einstein was the first person to realize that empty space is not nothing. Space has amazing properties, many of which are just beginning to be understood. The first property that Einstein discovered is that it is possible for more space to come into existence.

Then one version of Einstein's gravity theory, the version that contains a cosmological constant, makes a second prediction: "empty space" can possess its own energy. Because this energy is a property of space itself, it would not be diluted as space expands. As more space comes into existence, more of this energy-of-space would appear. As a result, this form of energy would cause the Universe to expand faster and faster.

Unfortunately, no one understands why the cosmological constant should even be there, much less why it would have exactly the right value to cause the observed acceleration of the Universe."

Dark Matter Core Defies Explanation

"This image shows the distribution of dark matter, galaxies, and hot gas in the core of the merging galaxy cluster Abell 520. The result could present a challenge to basic theories of dark matter.

Another explanation for how space acquires energy comes from the quantum theory of matter. In this theory, "empty space" is actually full of temporary ("virtual") particles that continually form and then disappear. But when physicists tried to calculate how much energy this would give empty space, the answer came out wrong - wrong by a lot. The number came out 10120 times too big. That's a 1 with 120 zeros after it. It's hard to get an answer that bad. So the mystery continues.

Another explanation for dark energy is that it is a new kind of dynamical energy fluid or field, something that fills all of space but something whose effect on the expansion of the Universe is the opposite of that of matter and normal energy. Some theorists have named this "quintessence," after the fifth element of the Greek philosophers. But, if quintessence is the answer, we still don't know what it is like, what it interacts with, or why it exists. So the mystery continues.

The last possibility is that Einstein's theory of gravity is not correct. That would not only affect the expansion of the Universe, but it would also affect the way that normal matter in galaxies and clusters of galaxies behaved. This fact would provide a way to decide if the solution to the dark energy problem is a new gravity theory or not: we could observe how galaxies come together in clusters.

But if it does turn out that a new theory of gravity is needed, what kind of theory would it be? How could it correctly describe the motion of the bodies in the Solar System, as Einstein's theory is known to do, and still give us the different prediction for the Universe that we need? There are candidate theories, but none are compelling. So the mystery continues.

The thing that is needed to decide between dark energy possibilities - a property of space, a new dynamic fluid, or a new theory of gravity - is more data, better data." (Quote continues next page.)

WHAT IS DARK MATTER?

Abell 2744: Pandora's Cluster Revealed

"One of the most complicated and dramatic collisions between galaxy clusters ever seen is captured in this new composite image of Abell 2744. The haze shows a map of the total mass concentration (mostly dark matter).

By fitting a theoretical model of the composition of the Universe to the combined set of cosmological observations, scientists have come up with the composition that we described above, ~68% dark energy, ~27% dark matter, ~5% normal matter. What is dark matter?

We are much more certain what dark matter is not than we are what it is. First, it is dark, meaning that it is not in the form of stars and planets that we see. Observations show that there is far too little visible matter in the Universe to make up the 27% required by the observations. Second, it is not in the form of dark clouds of normal matter — matter made up of particles called baryons.

We know this because we would be able to detect baryonic clouds by their absorption of radiation passing through them. Third, dark matter is not antimatter, because we do not see the unique gamma rays that are produced when antimatter annihilates with matter. Finally, we can rule out large galaxy-sized black holes on the basis of how many gravitational lenses we see. High concentrations of matter bend light passing near them from objects further away, but we do not see enough lensing events to suggest that such objects make up the required 25% dark matter contribution.

However, at this point, there are still a few dark matter possibilities that are viable. Baryonic matter could still make up the dark matter if it were all tied up in brown dwarfs or in small, dense chunks of heavy elements. These possibilities are known as massive compact halo objects, or "MACHOs". But the most common view is that dark matter is not baryonic at all, but that it is made up of other, more exotic particles like axions or WIMPS (Weakly Interacting Massive Particles)." End of NASA quote: See: science.nasa.gov/astrophysics/focus-areas/what-is-dark-energy 2

Author's Hypothesis

As noted in the first part of this chapter referencing the controversy between famed physicists Dr. Hawkins and Dr. Thorne vs Dr. John Preskill, information is not lost in the infamous black holes we've heard about. Upon death, when our energy leaves this dimension of light and is swept into another, the information we had is not absorbed into that hole. It is sent back out to space (at least in part) in the form of radiation energy (Dark Matter) and is part of what has been known since antiquity. Therefore nothing is new. Could it be used by us again? Could Dark Energy or Dark Matter actually be storing information from antiquity?

> This idea of mine seems to be corroborated in the Bible: Ecclesiastes 1:10:) "Is there anything whereof it may be said, "See', this is new? It hath been already of old time, which was before us."

Chapter 8
Language and Bigfoot

Continuing research increasingly supports that humans, dolphins, and apes evolved from a common ancestor. Dolphins and whales used to live on land and new research shows that humans are more closely related to dolphins than they are related to apes.

There is much research and many published papers regarding the highly developed language of dolphins. Basically, dolphins comprehend elements of human language, as well as having a complex visual language of their own. Kassewitz states, in a peer reviewed paper, that dolphins also have the unique ability to transfer sound signals into pictures in their brain.

Jack Kassewitz, Journal of Marine Science: Research & Development, August 2016. Also at www.speakdolphin.com

Joan Ocean studies the cognizance of dolphins. Ms. Ocean, Master of Science, is internationally known, and has published several books and audio DVDs on her work in the field of human-dolphin, and whale communication. Through quirks of fate, her life and study of marine mammals, especially a pod of 200 Hawaiian Spinner dolphins, eventually turned her to Bigfoot and Bigfoot communication. You will read more about Joan Ocean in Chapter 20.

> "We now know that some, if not all, Bigfoots have the ability to speak using a complex language." R. Scott Nelson, retired Navy crypto-linguist:

Language and Bigfoot

From my *Voices in the Wilderness*, 2012, self-published, Mariposa: "… It finally happened in March 2008 when Scott Nelson contacted me. He is retired from Naval Intelligence as a Crypto-Linguist and teaches Russian, Spanish, and Persian at a college in Missouri. He asked if he could listen to, and perhaps, study, the original recordings."

This brief introduction resulted in years of work with Scott. During that time I learned the basics of language and its importance to human development and society.

My thoughts regarding language are not unique in advanced circles of Bigfoot research, but I notice that some people seem very passive and complacent with this subject like it's no big deal. Our awareness is that only humans are known to converse with an actual sapient language, so it is important to understand the difference between sentience and/or cognizance and sapience.

Many different animals are sentient or have the ability to be self-aware. However only humans are considered sapient; or have the ability to reason with abstract thought, remove themselves from the present; and, relay thoughts from the present, past, and speculative future through spoken language.

> "New studies show that our language ability has a unique effect on how things turn out, or in short, affect the future. Prior to a spoken word from a human's mouth, the thought has occurred to that persons' mind. What happens after that is remarkable."
> Lera Boroditsky.Edge.org

Language and Bigfoot

Do we have the ability to actually change anything with our words — could we actually be subject to whatever comes out of our mouth?

"Death and life are in the power of the tongue, and those who love it will eat its fruits." (Proverbs 18:21):

"In the Beginning was the Word, and the Word was with God and the Word was God," (John 1:1)).

These passages lead us to believe God created everything through words (vibrational waves), and thoughts. So, are we not made in His image? I say, "Yes we are" and that makes us creators.

Unlike other animals on earth, the Bigfoots that I interacted with have a complex language and seem to have the ability to reason with wisdom and discernment.

Many times over the years I've been asked to listen to purported sound recordings of what some think for sure are sounds representing Bigfoot language. And many times I've said that I'm no expert with animal sounds or even language, as far as linguistics is concerned. However, after spending much time with crypto-linguist Scott Nelson, I feel I know more about what is needed to determine if language is within unusual sounds or vocalizations that come my way.

While hunting in the Sierra Nevada Mountains, I and my buddies had an unusual number of close encounters with the giants. Eventually the six of us made a substantial number of cassette recordings. One of us, Al Berry who is profiled in Chapter 19, was an investigative reporter.

In 1977 Al was able to engage Dr. R. Lynn Kirlin, Electrical Sound Engineer at the University of Wyoming, to analyze our recordings. Dr. Kirlin assessed the vocalizations and determined that they were made at the time of the recordings with no speed alterations or manipulations. (See Chapter 12, Professional Findings). Unbeknown to Dr. Kirlin, years earlier in 1973 Syntonic Research Laboratory came to the same initial conclusion as made by Kirlin after a year-long study.

Prior to engaging Dr. Kirlin, in 1973 Al engaged Syntonic Research Laboratory to analyze the tapes. (Syntonic Research was responsible for the study of the Nixon Watergate tapes.) Since their cost was prohibitive for Al, Syntonic did not do a complete analysis, but with professional optimism, offered an initial statement the tapes were genuine and not manipulated.

"The vocalizations had occurred at the time of the original recordings, were spontaneous, and were too powerful to have been human-made." From a report by I.E. Teibel, president of Syntonic Research, Inc., New York, 1973. More on Mr. Teibel's analysis is presented in Chapter 19.

Although we figured these creatures were jabbering amongst themselves with some type of primitive and very unusual verbal communication it was not until after I met Scott Nelson in 2008 that I discovered our mountain beings were using a complex language. And, I didn't know a man with language skills like Scott's existed; someone who could actually transcribe a language from an unknown source. Read more about Mr. Nelson in Chapter 12, "Professional Findings."

R. Scott Nelson

So what is it that actually makes up the human definition of language? In linguistics, a morpheme stream is a combination of words (or phonemes) that makes up a phonetic unit in a language. A morpheme is not identical to a word, and the principal difference between the two is that a morpheme may or may not stand alone, whereas a word, by definition, is freestanding.

What is needed to determine if a recorded sound may represent Bigfoot language? First, establish i f the sounds have a 'morpheme stream' or not — a sentence made up of ten to twelve words. Second, if recordings have a 60 cycle hum, it would indicate that the sounds were pre-recorded in a studio and would disqualify the recordings. A qualified sound engineer could assess this, and could also find out if the sounds on the recordings had been manipulated in any way, e.g., slowed down or sped up. If the sounds pass this critique they were probably made by a human or by a Bigfoot.

Language and Bigfoot

The Russians seem to have studied speech and its implications more deeply than we have in the United States. I journeyed to Russia in October 2011 and spent time there with two noted scientists, Dr. Igor Burtsev and Dr. Dmitri Bayanov. I also had the honor of speaking at the State Darwin Museum of Science in Moscow with Dr. Jeff Meldrum (United States) and Dr. John Bindernagel (Canada). Afterward, we all went to Siberia on a trip to explore possible Yeti habitat.

Dr. Jeff Meldrum, Ron Morehead, Russian interpreter, and Dr. John Bindernagel in remote Siberia.

Russian DNA Research

"These groundbreaking, amazing experiments with world-changing implications show that language, sound and light frequencies have a far more powerful influence on our DNA than was ever imagined." Dr.Vladimir Poponin, Russia. Dr. Poponin is a quantum physicist and leading expert in quantum biology.

Circa 2010 Russian biophysicist and molecular biologist Pjotr Garjajev and his colleagues have been carrying out cutting-edge research on the more esoteric natures of DNA. From the German book *Vernetzte Intelligenz* by Grazyna Fosar and Franz Bludorf (summarized and translated by Baerbel: 2001):

"The latest research explains phenomena such as clairvoyance, intuition, spontaneous and remote acts of healing, self-healing, affirmation techniques, unusual light-auras around people (namely spiritual masters), mind's influence on weather-patterns and much more. The Russian scientists also found out that our DNA can cause disturbing patterns in the vacuum, thus producing magnetized wormholes! Wormholes are the microscopic equivalents of the so-called Einstein-Rosen bridges in the vicinity of black holes (left by burned-out stars). These are tunnel connections between entirely different areas in the universe through which information can be transmitted outside of space and time. The DNA attracts these bits of information and passes them on to our consciousness...

Russian researcher Dr.Vladimir Poponin put DNA in a tube and beamed a laser through it. When the DNA was removed, the laser light continued spiraling on its own, like it would through a crystal! This effect is called 'Phantom DNA Effect'.

It is surmised that energy from outside of space and time still flows through the activated wormholes after the DNA was removed. The side effect encountered most often in hyper-communication and also in human beings are inexplicable electromagnetic fields in the vicinity of the persons concerned. Electronic devices like CD players and the like can be irritated and cease to function for hours. When the electromagnetic field slowly dissipates, the devices function normally again. Many healers and psychics know this effect from their work.

The most astonishing experiment that was performed by Garjajev's group is the reprogramming of the DNA codon sequences using modulated laser light. From their discovered grammatical syntax of the DNA language, they were able to modulate coherent laser light and even radio waves and add semantics (meaning) to the carrier wave. In this way, they were able to reprogram in vivo DNA in living organisms, by using the correct resonant frequencies of DNA. The most impressive discovery made so far is that spoken language can be modulated to the carrier wave with the same reprogramming effect. Now, this is a baffling and stunning scientific discovery! Our own DNA can simply be reprogrammed by human speech, supposing that the words are modulated on the correct carrier frequencies!

Whereas western science uses complicated biochemical processes to cut and paste DNA triplets in the DNA molecule, Russian scientists use modulated laser light to do exactly the same thing. The Russians have proven to be very successful in repairing damaged DNA material in vivo!

Laser light therapies based on Garjajev's findings are already applied in some European academic hospitals with success on various sorts of skin cancer. The cancer is cured without any remaining scars."

Source: www.noeticdigest.wordpress.com

So, how significant is language? Humans have language; and, it gave us dominance over everything on this earth. Words can create, so let's use it for good (Genesis 1:26)). Say what you want, not what you don't want. Speak positively, or don't speak at all. Conceive the idea in your mind, feel it in your soul (heart), speak it over and over. After that, the universe is at work and nothing can stop it. And, it's not our job to tell the universe 'how' or 'when' to make something come about. Also, it's important to understand that the brain is the servant of the heart — "feel it."

The author in north central Siberia.
Photograph taken by interpreter.

Chapter 9
Quantum Entanglement

"Scientists are closing in on definitively proving "spooky action at a distance" — the notion that entangled particles can instantly communicate. What Einstein called "spooky action at a distance" links pairs of particles even when separated." www.livescience.com: Strange News

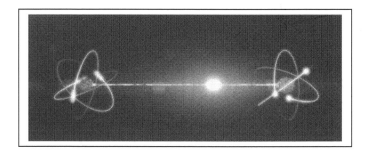

As you read this chapter you may think that it doesn't exactly fit in a Bigfoot book. But honestly it is at the core of an enlightened human experience and that is my point exactly. An experience that I believe many of the Sasquatch people live in daily and the place from which they operate.

What one animal knows, the others know? Take our dogs for instance…if the little one needs to go out or is hungry the big one barks for her. What about birds flocking together, moving in unison, does the lead bird say, "Let's all turn left when I flap my right wing twice? No. The same applies to a school of fish…they don't talk, they communicate with a collective conscience. Now, I'm well aware that there are studies of this behavior that are labeled differently, but they obviously have left many questions unanswered that are easily answered in the Quantum Entanglement realm.

Most of us humans are too fear based and protective to relax into the collective conscience — the Oneness that Christ spoke of in John 17:21.

How does this pertain to Bigfoot you might ask?

"I have noticed a pattern in experiential reports with the Sasquatch... If with every cell, we are seeking love, then we receive a feeling of deep love. If we are seeking fun and mischief then we get beings who play jokes with us, and if due to our interpretation of our life circumstances we have not moved beyond feeling love as ego recognition, then that is what we receive...findings and interactions that make us feel important....and loved...not realizing that what we are getting is an exact reflection of our mental/emotional psyche. This may be really hard to swallow, but if they are indeed trying to teach us and if some of our human idioms run true and we can stand to look long enough...then a mirror is indeed our most powerful teacher."-Keri Campbell

The Russians have studied this subject intensely and I find the following article written in 2007 by two German authors (Grazyna Gosar and Franz Bludorf) very fascinating. Their book is *Vernetzte Intelligenz*, meaning 'Networked Intelligence.'

From *Vernetzte Intelligenz* by von Grazyna Fosar und Franz Bludorf: ISBN 3930243237. For partial English language excerpts please go to: The Global Consciousness Project: noosphere.princeton.edu/gcpintro.html. As of 2017 this book is currently only available in German. Author contact: www.fosar-bludorf.com

"The human DNA is a biological Internet and superior in many aspects to the artificial one. Russian scientific research directly or indirectly explains phenomena such as clairvoyance, intuition, spontaneous and remote acts of healing, self-healing, affirmation techniques, unusual light/auras around people (namely spiritual masters), mind's influence on weather patterns and much more. In addition, there is evidence for a whole new type of medicine in which DNA can be influenced and reprogrammed by words and frequencies WITHOUT cutting out and replacing single genes.

Only 10% of our DNA is being used for building proteins. It is this subset of DNA that is of interest to western researchers and is being examined and categorized. The other 90% are considered "junk DNA." The Russian researchers, however, convinced that nature was not dumb, joined linguists and geneticists in a venture to explore those 90% of "junk DNA."

Their results, findings and conclusions are simply revolutionary! According to them, our DNA is not only responsible for the construction of our body but also serves as data storage and in communication. The Russian linguists found that the genetic code, especially in the apparently useless 90%, follows the same rules as all our human languages.

To this end they compared the rules of syntax (the way in which words are put together to form phrases and sentences), semantics (the study of meaning in language forms) and the basic rules of grammar. They found that the alkalines of our DNA follow a regular grammar and do have set rules just like our languages. So human languages did not appear coincidentally but are a reflection of our inherent DNA.

The Russian biophysicist and molecular biologist Pjotr Garjajev and his colleagues also explored the vibrational behavior of the DNA. The bottom line was: "Living chromosomes function just like solitonic/holographic computers using the endogenous DNA laser radiation." This means that they managed for example to modulate certain frequency patterns onto a laser ray and with it influenced the DNA frequency and thus the genetic information itself. Since the basic structure of DNA-alkaline pairs and of language (as explained earlier) are of the same structure, no DNA decoding is necessary.

One can simply use words and sentences of the human language! This, too, was experimentally proven! Living DNA substance (in living tissue, not in vitro) will always react to language-modulated laser rays and even to radio waves, if the proper frequencies are being used.

This finally and scientifically explains why affirmations, autogenous training, hypnosis and the like can have such strong effects on humans and their bodies. It is entirely normal and natural for our DNA to react to language. While western researchers cut single genes from the DNA strands and insert them elsewhere, the Russians enthusiastically worked on devices that can influence the cellular metabolism through suitable modulated radio and light frequencies and thus repair genetic defects.

Garjajev's research group succeeded in proving that with this method chromosomes damaged by x-rays for example can be repaired. They even captured information patterns of a particular DNA and transmitted it onto another, thus reprogramming cells to another genome?

So, they successfully transformed, for example, frog embryos to salamander embryos simply by transmitting the DNA information patterns! This way the entire information was transmitted without any of the side effects or disharmonies encountered when cutting out and re-introducing single genes from the DNA.

This represents an unbelievable, world-transforming revolution and sensation! All this by simply applying vibration and language instead of the archaic cutting-out procedure! This experiment points to the immense power of wave genetics, which obviously has a greater influence on the formation of organisms than the biochemical processes of alkaline sequences.

Esoteric and spiritual teachers have known for ages that our body is programmable by language, words and thought. This has now been scientifically proven and explained.

Of course the frequency has to be correct. And this is why not everybody is equally successful or can do it with always the same strength. The individual person must work on the inner processes and maturity in order to establish a conscious communication with the DNA. The Russian researchers work on a method that is not dependent on these factors but will ALWAYS work, provided one uses the correct frequency.

But the higher developed an individual's consciousness is, the less need is there for any type of device! One can achieve these results by oneself, and science will finally stop to laugh at such ideas and will confirm and explain the results. And it doesn't end there.

The Russian scientists also found out that our DNA can cause disturbing patterns in the vacuum, thus producing magnetized wormholes! Wormholes are the microscopic equivalents of the so-called Einstein-Rosen bridges in the vicinity of black holes left by burned-out stars. These are tunnel connections between entirely different areas in the universe through which information can be transmitted outside of space and time. The DNA attracts these bits of information and passes them on to our consciousness.

This process of hyper communication is most effective in a state of relaxation. Stress, worries or a hyperactive intellect prevent successful hyper communication or the information will be totally distorted and useless.

In nature, hyper communication has been successfully applied for millions of years. The organized flow of life in insect states proves this dramatically.

Modern man knows it only on a much more subtle level as "intuition." But we, too, can regain full use of it.

An example from Nature: When a queen ant is spatially separated from her colony, building still continues fervently and according to plan. If the queen is killed, however, all work in the colony stops. No ant knows what to do. Apparently the queen sends the "building plans" also from far away via the group consciousness of her subjects. She can be as far away as she wants, as long as she is alive.

In man, hyper communication is most often encountered when one suddenly gains access to information that is outside one's knowledge base. Such hyper communication is then experienced as inspiration or intuition. The Italian composer Giuseppe Tartini for instance dreamt one night that a devil sat at his bedside playing the violin. The next morning Tartini was able to note down the piece exactly from memory, he called it the Devil's Trill Sonata.

For years, a 42-year old male nurse dreamt of a situation in which he was hooked up to a kind of knowledge CD-ROM. Verifiable knowledge from all imaginable fields was then transmitted to him that he was able to recall in the morning. There was such a flood of information that it seemed a whole encyclopedia was transmitted at night. The majority of facts was outside his personal knowledge base and reached technical details about which he knew absolutely nothing.

When hyper communication occurs, one can observe in the DNA as well as in the human being special phenomena.

The Russian scientists irradiated DNA samples with laser light. On screen a typical wave pattern was formed. When they removed the DNA sample, the wave pattern did not disappear, it remained. Many control experiments showed that the pattern still came from the removed sample, whose energy field apparently remained by itself. This effect is now called phantom DNA effect.

It is surmised that energy from outside of space and time still flows through the activated wormholes after the DNA was removed. The side effects encountered most often in hyper communication also in human beings are inexplicable electromagnetic fields in the vicinity of the persons concerned. Electronic devices like CD players and the like can be irradiated and cease to function for hours. When the electromagnetic field slowly dissipates, the devices function normally again.

Many healers and psychics know this effect from their work. The better the atmosphere and the energy, the more frustrating it is that the recording device stops functioning and recording exactly at that moment. And repeated switching on and off after the session does not restore function yet, but next morning all is back to normal. Perhaps this is reassuring to read for many, as it has nothing to do with them being technically inept; it means they are good at hyper communication.

Grazyna Gosar and Franz Bludorf explain these connections precisely and clearly. The authors also quote sources presuming that in earlier times humanity had been, just like the animals, very strongly connected to the group consciousness and acted as a group.

To develop and experience individuality we humans however had to forget hyper communication almost completely. Now that we are fairly stable in our individual consciousness, we can create a new form of group consciousness, namely one, in which we attain access to all information via our DNA without being forced or remotely controlled about what to do with that information.

We now know that just as on the internet our DNA can feed its proper data into the network, can call up data from the network and can establish contact with other participants in the network. Remote healing, telepathy or "remote sensing" about the state of relatives, etc. can thus be explained.

Some animals also know from afar when their owners plan to return home. That can be freshly interpreted and explained via the concepts of group consciousness and hyper communication. Any collective consciousness cannot be sensibly used over any period of time without a distinctive individuality. Otherwise we would revert to a primitive herd instinct that is easily manipulated.

Hyper communication in the new millennium means something quite different:

Researchers think that if humans with full individuality would regain group consciousness, they would have a god-like power to create, alter and shape things on Earth! AND humanity is collectively moving toward such a group consciousness of the new kind.

Fifty percent of today's children will be problem children as soon as they go to school. The system lumps everyone together and demands adjustment. But the individuality of today's children is so strong that that they refuse this adjustment and giving up their idiosyncrasies in the most diverse ways.

At the same time more and more clairvoyant children are born [see the book *China's Indigo Children*, by Paul Dong or the chapter about Indigos in the book *Nutze die taeglichen Wunder* "Make Use of the Daily Wonders." Something in those children is striving more and more towards the group consciousness of the new kind, and it will no longer be suppressed. As a rule, weather for example is rather difficult to influence by a single individual. But it may be influenced by a group consciousness (nothing new to some tribes doing it in their rain dances).

Weather is strongly influenced by Earth resonance frequencies, the so-called Schumann frequencies. But those same frequencies are also produced in our brains, and when many people synchronize their thinking or individuals (spiritual masters, for instance) focus their thoughts in a laser-like fashion, then it is scientifically speaking not at all surprising if they can thus influence weather. Baerbel excerpted, translated and commented on *Vernetzte Intelligenz*

Researchers in group consciousness have formulated the theory of **Type I civilizations**. A humanity that developed a group consciousness of the new kind would have neither environmental problems nor scarcity of energy. For if it were to use its mental power as a unified civilization, it would have control of the energies of its home planet as a natural consequence. And that includes all natural catastrophes!!!

A theoretical **Type II civilization** would even be able to control all energies of their home galaxy. In the book *Nutze die taeglichen Wunder*, the author has described an example of this:

Whenever a great many people focus their attention or consciousness on something similar like Christmas time, football world championship or the funeral of Lady Diana in England then certain random number generators in computers start to deliver ordered numbers instead of the random ones. An ordered group consciousness creates order in its whole surroundings!

When a great number of people get together very closely, potentials of violence also dissolve. It looks as if here, too, a kind of humanitarian consciousness of all humanity is created."

"It apparently is also an organic superconductor that can work at normal body temperature. Artificial superconductors require extremely low temperatures of between 200 and 140°C to function. As one recently learned, all superconductors are able to store light and thus information. This is a further explanation of how the DNA can store information. There is another phenomenon linked to DNA and wormholes. Normally, these supersmall wormholes are highly unstable and are maintained only for the tiniest fractions of a second. Under certain conditions stable wormholes can organize themselves which then form distinctive vacuum domains in which for example gravity can transform into electricity.

Vacuum domains are self-radiant balls of ionized gas that contain considerable amounts of energy. There are regions in Russia where such radiant balls appear very often. Following the ensuing confusion the Russians started massive research programs leading finally to some of the discoveries mentioned above. Many people know vacuum domains as shiny balls in the sky.

The attentive look at them in wonder and ask themselves, what they could be? I thought once: "Hello up there. If you happen to be a UFO, fly in a triangle." And suddenly, the light balls moved in a triangle. Or they shot across the sky like ice hockey pucks. They accelerated from zero to crazy speeds while sliding gently across the sky. One is left gawking. Myself have, as many others, too, thought them to be UFOs. Friendly ones, apparently, as they flew in triangles just to please.

Now the Russians found in the regions, where vacuum domains appear often, that sometimes balls of light fly from the ground upwards into the sky; and, that

these balls can be guided by thought. One has found out that vacuum domains emit waves of low frequency as they are also produced in our brains.

And because of this similarity of waves they are able to react to our thoughts. To run excitedly into one that is on ground level might not be such a great idea, because those balls of light can contain immense energies and are able to mutate our genes. They can, they don't necessarily have to, one has to say. For many spiritual teachers also produce such visible balls or columns of light in deep meditation or during energy work which trigger decidedly pleasant feelings and do not cause any harm. Apparently this is also dependent on some inner order and on the quality and provenance of the vacuum domain. There are some spiritual teachers (the young Englishman Ananda, for example) with whom nothing is seen at first, but when one tries to take a photograph while they sit and speak or meditate in hyper communication, one gets only a picture of a white cloud on a chair.

In some Earth healing projects such light effects also appear on photographs. Simply put, these phenomena have to do with gravity and anti-gravity forces that are also exactly described in the book and with ever more stable wormholes and hyper communication and thus with energies from outside our time and space structure.

Earlier generations that got in contact with such hyper communication experiences and visible vacuum domains were convinced that an angel had appeared before them. And we cannot be too sure to what forms of consciousness we can get access when using hyper communication. Not having scientific proof for their actual existence (people having had such experiences do NOT all suffer from hallucinations) does not mean that there is no metaphysical background to it. We have simply made another giant step towards understanding our reality.

Official science also knows of gravity anomalies on Earth (that contribute to the formation of vacuum domains), but only of ones below one percent. But recently gravity anomalies have been found of between three and four percent. One of these places is Rocca di Papa, south of Rome (exact location in the book *Vernetzte Intelligenz* plus several others). Round objects of all kinds, from balls to full buses, roll uphill. But the stretch in Rocca di Papa is rather short, and defying logic skeptics still flee to the theory of optical illusion (which it cannot be due to several features of the location)." End of quote.

Author's Note:

In my opinion this article lends credence in a big way to The Quantum Bigfoot. It speaks of multiple minds on the same vibrational frequency, working to create a shared reality.

If you read my first book, *Voices in The Wilderness*, you may remember Warren and I being stopped in our tracks by some sort of force field. We were headed up a slope in the direction of 'something" that sounded just like a Bigfoot. We couldn't see it but we could certainly hear it. We got to a point that was about 15' away from the large tree that we thought it was behind and were suddenly stopped in our tracks. This same thing happened to me on another occasion when I was alone. This force of energy, kept us/me from being able to move any further forward. And it was not because of fear. By this point in time we assumed they meant us no harm.

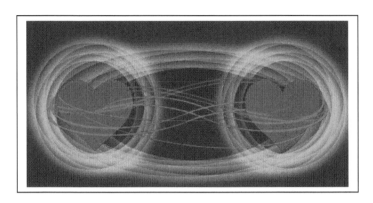

Quantum Entanglement at its Best

In case you don't believe any of this: Years ago when I became single, I wished for a woman in my life that could make me a better man. Because I wasn't ready for the front porch rocker, I asked for a woman that was younger than me, smarter than me, and vibrant. Because I had taken Yoga in my younger days, I asked for a woman that is proficient in Yoga and could actually take care of herself in case I wasn't around.

'Pretty is as pretty does' is all good, but I wanted 'pretty' on the inside as well as a good person all-around. So I asked for someone beautiful inside and out who was also spiritually compatible.

Although I believe in life after death (somewhere), the alternative for not getting older was not a cherished thought; there was a lot more for me to do.

So because I had no desire whatsoever for my fun years to be behind me, I asked for a woman who was a dietician and perhaps a nutritionist too. Good healthy eating habits, that's what I wanted.

Plus she had to be someone that would eagerly and happily join me on my Bigfoot adventures. But the most important thing I asked for was a lady — a real lady, one that doesn't mind getting her hands dirty occasionally. I thought if I could make things happen with my words and thoughts I may as well go for the perfect 'Holy Grail' woman.

Who did I ask and who did I get? I asked the Universe (God) and I got Keri. She is the perfect woman for me…everything I asked for and more. She's just as beautiful on the inside as on the outside; always wanting to help people and completely honest, all the time. She's a Yoga Instructor, a certified raw/superfood counselor, a massage therapist, Reiki practitioner, energy healer, and now a Bigfoot enthusiast.

Having taken Shotokan karate I know she can protect herself (and me). She drives when I get sleepy; she's an excellent cook, and because she was previously educated as an English major she is the perfect one for looking over my writings — everything a man could want and more — perfectly compatible.

In the past, I've been very independent, often a bit wild, and at times almost misogynistic. Every day she tries very hard to make a gentleman out of me and I think (hope) it's beginning to work. But occasionally I still fall short.

I thank the Universe for my wonderful life and for the woman I asked for — my wonderful beloved Keri — the embodiment of pure Love.

Ron Morehead, 2017

Chapter 10
State of Mind — A Bigfoot Connection?

In my first book, *Voices in the Wilderness*, I mentioned my encounter during 2011. I was at our *Sierra Camp*, reading a book when an almost deafening wood knock exploded very close to me — it was still daylight and I was totally awake. I was inside a small tent which I erected because our shelter was still suffering from the winter snow load and the mosquitoes were horrible.

Scott Nelson (Crypto-Linguist) and I had been there three times already that summer and nothing definitive had taken place. After Scott left for his teaching job in Missouri I wanted to go back again; I had to know if the creatures were still around or not. I prepared for a three-day stay, but on this fourth trip that summer, I went in alone. Suffice it to say, things got a little exciting after dark.

What I didn't say in my book was that I was totally relaxed — both when the wood pop occurred and later that night when something bi-pedal was just outside my tent slowly walking around. It was at 10 p.m. and I was in the Alpha state of relaxation (the sleep stage where we're not totally asleep, but not totally awake). Bigfoot seems to be more active during the night time when our body is at rest and our mind is not occupied with problems, e.g., "did I bring enough food, will the employees show up tomorrow, why am I alone up here? Etc."

Looking back on all the events that have taken place at this camp I realized that all of us were in a relaxed, contented, state most of the time. We were all away from our jobs and businesses and were enjoying the pristine area. I think our attitude may have something to do with some of the encounters. When in that vibrational state, the frequency of our bodies is different and it might be a better time for these beings to interact and try to communicate with us.

State of Mind – A Bigfoot Connection?

Although I talked about vibrations in Chapter 6, this chapter will expand on it and also offer my take on how important attitude and relaxation can be. I think all these factors play a role in helping the field researchers have a better experience.

Human Vibrational Frequency

If the two hemispheres of our brain are synchronized with each other at 8Hz, they work more harmoniously and with a maximum flow of information. In other words, the frequency of 8Hz seems to be the key to the full and sovereign activation potential of our brain. Attunedvibrations.com

Optimal frequency is when all is in order - when we are in balance, our vibration is in unison with the law of vibration - when we are in tune and each cell of our body vibrates at the frequency it was designed to... there is no effort, no conflict. naturalhealthzone.com

The Earth's Vibrational Frequency

The heartbeat of the Earth is better known as Schumann resonance and is named after physicist Winfried Otto Schumann, who documented it mathematically in 1952. It is said that 8Hz is the fundamental "beat" of the planet. Schumann's resonance is a global electromagnetic resonance, which has its origin in electrical discharges of lightning within the cavity existing between the Earth's surface and the ionosphere. This cavity resonates with electromagnetic waves in the extremely low frequencies of approximately 7.86Hz – 8Hz. Source: theeventchronical.com

State of Mind – A Bigfoot Connection?

Author's note: Human optimal vibrational frequency is around 8Hz and the earth's frequency is about the same. Could it be that Bigfoot is in tune with the earth and during the alpha state we are more closely in tune with Bigfoot? Is this 8Hz fundamental harmonious frequency why interactions become more likely?

Sleep

"It is important to know that all humans display five different types of electrical patterns or "brain waves" across the cortex. The brain waves can be observed with an EEG, an "electroencephalograph," a tool that allows researchers to note brain wave patterns. Each brain wave has a purpose and helps serve us in optimal mental functioning.

Our brain's ability to become flexible and/or transition through various brain wave frequencies plays a large role in how successful we are at managing stress, focusing on tasks, and getting a good night's sleep. If one of the five types of brain waves is either overproduced and/or under produced in our brain, it can cause problems. For this reason, it is important to understand that there is no single brain wave that is "better" or more "optimal" than the others.

Each serves a purpose to help us cope with various situations – whether it is to help us process and learn new information or help us calm down after a long stressful day. The five brain waves in order of highest frequency to lowest are as follows: gamma, beta, alpha,

State of Mind – A Bigfoot Connection?

theta, and delta." Source: Mental Health Blog Daily

"Scientists have assigned names to four frequency ranges of waves that can be distinguished in an EEG trace. From the highest to the lowest frequency, these waves are as follows.

- Beta waves: have a frequency range from 13-15 to 60 Hz and an amplitude of about 30 μV. Beta waves are the ones registered on an EEG when the subject is awake, alert, and actively processing information. Some scientists distinguish the range above 30-35 Hz as gamma waves, which may be related to consciousness–that is, the making of connections among various parts of the brain in order to form coherent concepts.

- Alpha waves: have a frequency range from eight to twelve Hz and an amplitude of 30 to 50 μV. Alpha waves are typically found in people who are awake but have their eyes closed and are relaxing or meditating.

- Theta waves have a frequency range from three to eight Hz and an amplitude of 50 to 100 μV. Theta waves are associated with memory, emotions, and activity in the limbic system.

State of Mind – A Bigfoot Connection?

- Delta waves: range from 0.5 to three or four Hz in frequency and 100 to 200 µV in amplitude. Delta waves are observed when individuals are in deep sleep or in a coma.

- Lastly, when there are no brain waves present, the EEG shows a flat-line trace, which is a clinical sign of brain death." From: Canadian Institutes of Health Research

Next, in Chapter 11, we will discuss the pineal gland, and how important it is to get good sleep. Until the 1950s, most people thought of sleep as a passive, dormant part of our daily lives. We now know that our brains are very active during sleep. So, does good rest in the field, combined with the harmonious frequency of 8Hz attract Bigfoot to relaxed sleeping humans? Reports are rife of Bigfoots approaching or watching sleeping humans. That is fact.

Chapter 11
The Pineal Gland

Egyptian temple hieroglyph of "The Third Eye"

In case you don't remember what Albert Einstein's equation ($E=mc^2$) translates to it is E for energy, M for mass, and C^2 for the speed of light squared.

Obviously we need light to see. Although the Pineal Gland possesses a lens, cornea and retina, it does not require light. Yet by many, including the ancients it is considered the third eye.

> "No man has seen God at any time," John 1:18
>
> Jesus Said, "…Anyone who has seen me has seen the father" John 14:9
>
> "I am the light of the world. Whoever follows me will not walk in darkness, but will have the light of life," John 8:12

The Pineal Gland

"The human body is the temple of God. One who kindles the light of awareness within gets true light." Source: Ancient Hinduism, Rig-Veda

The list goes on and on, beginning with ancient scriptures that mention the light inside of us. Is all this Bible stuff for religion only, or could there actually be a physical way to connect to a Supreme Being — the God who we don't see; or to a possible higher dimension? A dimension where God normally hangs out and communicates to the people He created? If so, how do we know how to do this?

Again, I think we should know and understand who we are as humans to fully understand who or what Bigfoot might represent. If those beings are part human, we need to know how they came about. Do they have knowledge of the cosmos? What could be their belief? They probably don't hold church in the forest on Sunday or brush with Fluoride toothpaste, calcifying their Pineal Gland.

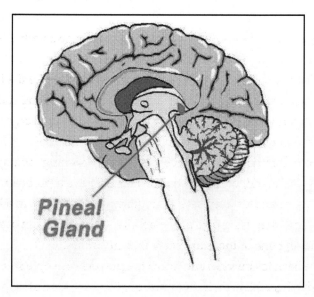

I'm not opposed to any religion, sect, or cult, *IF* its foundation follows, promotes and fully practices 'love' as its core belief. For much of my life, I've heard one religious denomination judge another — yet often both read the same Bible, that very same Bible, which tells them to not judge others. Just because one religion believes in multiple embodiments over eons to balance their karma and the other does not, does that mean that someone has to be all wrong? I say "yes" someone is probably wrong, but we are not supposed to judge.

> "Whoever undertakes to set himself up as a judge of Truth and Knowledge is shipwrecked by the laughter of the gods," Albert Einstein

For eons, civilizations have seen themselves as physical beings, and we are. I believe, like many others, that there is more to us than just a body — how about "Body, Soul, and Spirit?" But, like others, I've asked the question, "How do we actually connect with this God, who is supposed to exist somewhere?"

After we're all pushing up daisies, do we really like the idea of using a staff to forever herd sheep in the cosmos, or burning endlessly in hell if you've been bad? Who's supposed to be in charge of hell anyway if all the evil doers are to be consumed in a fire?

Is our connection to God just on Sunday morning, perhaps at a loved ones funeral, or maybe only when we have a problem in our life? This, in-and-of-itself says that many people believe in a higher power. I call that using God as a spare tire — bringing Him out when we're having one of those untimely flats in our life.

I've been to churches and heard the prayers of people when they ask God to heal them, their love ones, or give some kind of help. They are sincere, praying over and over in the name of Jesus. After all, He said in John 14:14, "Ask anything in My name and it shall be given you." They pray and still walk out not healed and their prayer not answered. What are they doing wrong? Could they not be connected to the source by which the healing comes from; an actual physical part of our body by which we can spiritually connect? What do the philosophers say?

> Author's note: If you really want to ask in "Jesus Name" I suggest doing it correctly. Jesus' actual name is Yeshua, it was being confused with Joshua, so the Greeks changed it. His name literally means "To Deliver or Rescue."

French philosopher René Descartes (1596-1650) emphasized the pineal gland in his writings, calling it the seat of the soul and "The part of the body in which the soul directly exercises its functions." He contended that this was the center where the soul and body interacted, and where we receive our messages from the Divine.

> Even Jesus says in Matthew 6:22: "The light of the body is the eye: if therefore thine eye be single, thy whole body shall be full of light," so the idea of there being a spiritual eye is not a new one.

The Concept of Temple

Model of Herod's Temple on the Mount

Now let's step back to an Old Testament account of the Temple that Solomon built. In the inner court of the temple, in the Holy of Holies, was the Ark of the Covenant. That was where the high priest would go once a year to offer atonement for the sins of the people. A veil (a very thick woven curtain) separated the Holy of Holies from the rest of the temple. When Jesus was crucified that heavy curtain was torn from top to bottom (Matthew 21:53). What made the tearing of that veil significant?

That temple was a physical metaphor for our body. I think the Holy of Holies relates directly to our pineal gland. When Christ was crucified the veil has been torn, there is now nothing in our way — we have access to the collective oneness that Christ mentioned in John 17:21, "That they all may be one; as thou, Father, art in me, and I in them, that they also may be one in us; that the world may believe that thou hast sent me." Our body is the temple of God and that's how and why God works through His created man (1 Corinthians 6:19, & 2 Corinthians 6:16).

Every human on this planet was made in the image of God. The communication with Him comes through understanding what Jesus taught — the oneness is open to those that recognize and use this! The key to real success is to learn how to listen to that entity within and follow that guide — responding to all issues with Love is what it's all about. Love from the heart, which is transferred to the brain, and spoken from the mouth. Remember prayer is when you talk to God and meditation is when God talks to you. "Be still, and know that I am God…" Psalms 46:10.

Next Kevan Ryon discusses the penal gland in "Awake and Empowered Expos:" Go to: awake andempoweredexpo.com

> "The Pineal Gland or sixth chakra is located deep in the center of the brain and was considered by French philosopher Rene Descartes (previously mentioned) to be the "seat of the soul." Mainstream society has long suppressed the science behind the true nature of the pineal, but evidence of its multifaceted function and symbolic history is coming to light as the consciousness of the planet rises. The pineal gland serves as a symbol of spirituality, intuition, and clairvoyance in many religions and societies across the globe.
>
> It is often referred to as the "third eye" because it possesses a lens, cornea, retina, and has light-sensitive internal structures that are similar to the rods and cones of the eyes. Once activated, it allows an individual to view the spiritual realms. Yet to the detriment of humanity, toxins in the environment and a loss of

spiritual knowledge have caused the third eye to calcify and become dormant in most people. However, an awakening is occurring that is allowing people to rediscover their third eye abilities as alternative science uncovers the once well-known mysteries of the sixth chakra."

Pineal Gland Function: Melatonin

The primary function of the pineal gland as understood by mainstream science is to produce melatonin during sleep, as well as regulate body temperature and skin coloration, all of which depend on the presence of environmental light. Melatonin is a hormone produced from serotonin which affects our mood. Melatonin is secreted by the pineal gland in the absence of light which is responsible for repairing cells, governing circadian rhythms, and eliminating free radicals. A deficiency can lead to insomnia and cancer and accelerate the aging process. Electromagnetic radiation and the presence of light during sleep also interfere with the signals that tell the pineal gland to start producing melatonin.

Canadian medical doctor, Roman Rozencwaig began studying the pineal gland and its ability to produce melatonin in the mid-1980's. He believes decreased melatonin production is a direct result of atrophy of the pineal gland, which in turn leads to aging and disease. As a person grows older, the cells of the pineal are damaged and calcified due to toxins in the environment and injury. This leads to decreased melatonin production and higher levels of serotonin, which may be responsible for cancer and other diseases.

Dr. Rozencwaig states that supplemental melatonin can drastically reduce the rate of pineal atrophy, slow down the aging process, and treat a wide array of diseases by boosting the immune system, allowing the pineal gland to maintain a balanced ratio between melatonin and serotonin levels in the body. He has treated several of his cancer patients with supplemental melatonin and many have seen a significant improvement in their condition. Even though melatonin is a vastly important part of pineal health, it is not the compound responsible for inducing mystical experiences.

"DMT or Di-methyl-tryptamine, is a naturally occurring psychedelic compound that is found throughout the body and produced by the pineal gland. DMT is believed to play a role in inducing mystical states of consciousness. DMT can be found in almost all plants and animals and has been thought of as the key to our spiritual gateway by many indigenous and ancient tribes, particularly those in South America.

In the mid-1800s Amazonian explorers described psychoactive powders and brews made from plants with high concentrations of DMT. The tribe members often recounted out of body experiences in which they would communicate with spirits and experience mystical visions. These tribes understood the nature of these substances far earlier than their western counterparts.

DMT was discovered in the blood, urine, tissue, and spinal fluid of humans in the mid-60s and became the first psychedelic substance to be produced internally by the body. However, DMT research encountered political opposition in the following decade due to the growing anti-psychedelic sentiment and was not studied again until the 90s."

In his book *DMT-The Spirit Molecule*, Dr. Rick Strassman discusses his groundbreaking research on DMT and its effects on the pineal gland. Dr. Strassman asserts that the pineal's location in the center of the brain makes it an ideal candidate for DMT production.

Since the third eye is close to the cerebrospinal fluid channels, this allows any secreted substance to be easily absorbed by the brain. Furthermore, the pineal is situated near the colliculi which are mounds of tissue that transmit emotional and sensory data to certain parts of the brain. It is also surrounded by the limbic system which governs the experience of emotions.

DMT is rapidly broken down by the body so the production center needs to be close to the emotional areas of the brain in order to have an effect on the consciousness of a person. The pineal contains the necessary enzymes to convert tryptamine (the base compound that derives melatonin and serotonin) into DMT.

It also produces compounds called beta-carbolines which prevent neurotransmitter inhibitor enzymes called monoamine oxidase (MAO) from breaking down DMT. However, the pineal has mechanisms to prevent spontaneous DMT production during normal activities.

For example, both the pineal and adrenal glands produce neurotransmitters called catecholamines. These are known to stimulate melatonin production in the pineal and are produced in response to stress in the adrenals. When the adrenal catecholamines are near the pineal its nerve cells rapidly get rid of them. The nerve cells act as a security system to prevent the release of DMT in periods of everyday stress.

Dr. Strassman postulates that large amounts DMT are released by the pineal at specific points in a person's life. "Pineal DMT production is the physical representation of non-material, or energetic, processes. It provides us with the vehicle to consciously experience the movement of our life-force in its most extreme manifestations."

He believes that DMT is released when the soul enters the fetus, at birth, during states of meditation or psychosis, and when the soul leaves the body. In patients with psychosis, enzymes that stimulate DMT production are elevated and MAO enzymes are decreased. Therefore, their third eye is unable to turn off DMT production, resulting in hallucinations. During states of deep meditation, people have reported having mystical experiences similar to those produced by psychedelics.

When a person enters a deep meditative state their brain waves become slower and more coherent as they reach a level of expanded awareness. Thus, the

The Pineal Gland

pineal vibrates at these same frequencies which trigger the release of DMT. (Source erowid.org/)

Author's note: Many encounters with Bigfoot at the Sierra Camp occur during the night when our vibrational frequency is probably low.

>Mr. Ryon continues: "During birth and death, Strassman speculates that large amounts of catecholamines are released. Since birth and death can be high-stress experiences, the pineal is overwhelmed with catecholamines which stimulate DMT release.
>
>Dr. Strassman and his research team conducted a study in which they gave several volunteers low, medium, and high doses of IV DMT, and measured its effects on the body. When volunteers were given a dose of .05mg/kg, no psychedelic effects were present, but many experienced feelings of joy and happiness.
>
>At a medium dose volunteers experienced rapid colorful visual patterns that were kaleidoscopic or geometric in nature. At higher doses, volunteers experienced complete loss of connection with reality as they were fully launched into a psychedelic experience. Many felt as if they had died, accompanied by feelings of bliss and euphoria, while a smaller percentage experienced feelings of terror and paranoia.
>
>Several reported traveling through realms with vibrant unfamiliar colors and shapes while whizzing past galaxies. Some volunteers even encountered beings who interacted with them and guided them through the DMT space. Overall the experiences were individual but similar in many respects. Strassman's research indicates that pineal DMT production may

play a vital role in inducing mystical experiences during near death experiences, psychotic states, and deep states of meditation.

Pineal Calcification and Activation

Even though the pineal gland contains chemicals and neurotransmitters to induce mystical experiences without the use of external psychedelics, the pineal glands of most people are calcified and dormant. Many of the toxins in our food and environment are responsible for pineal calcification. Calcification prevents the pineal from functioning correctly and leads to decreased melatonin and DMT production. Fluoride, mercury, and processed foods, are some of the substances that contribute to pineal calcification. Fluoride is found in toothpaste and most of America's water supply and is one of the chief causes of pineal calcification. The calcified pineal contains the highest concentration of fluoride in the body.

Fluoride and mercury have been shown to affect brain function and even lower IQ. Mercury can be found in vaccines and tooth fillings and has been linked to the onset of autism in young children. However, fluoride can be removed from tap water with filters specifically designed for fluoride removal. Mercury and other heavy metals can be removed from the body with the correct supplements.

Processed foods contain a plethora of additives, chemicals, and artificial compounds that indirectly cause pineal calcification, as well as a host of other health problems. Over 50% of the standard American diet consists of processed foods. Since their inception in the early 20s, processed foods have become a staple in many American diets, with Americans consuming 31% more processed food than fresh food.

There are numerous ways to activate and open the third eye, and each method can be individualized to suit the needs of the person. One of the most widely practiced pineal activation techniques is meditation. Dr. Joe Dispenza, author of *Evolve Your Brain: The Science of Changing Your Mind*, claims that through the science of neuroplasticity, a person can rewire their brain.

In his research with cases of spontaneous remission, Dr. Dispenza discovered that those who had achieved total healing from terminal illnesses were able to rewire their brain through positive intention and thought. When a person is awake and alert they are generally in a beta state.

When one begins to enter a meditative trance the brain enters alpha, theta, and gamma states, with the latter being the slowest and most conducive to healing. * These slow frequencies allow the brain to readily access the field of quantum possibility when an intention is set. This is because the brain is not preoccupied with tasks of daily survival and it is able to expand its focus beyond everyday needs. Therefore if a person sets an intention to open their third eye, it can be activated with regular meditation practice."

End of Kevan Ryon quote.

Author's note: If Bigfoot can operate in a quantum field, could our slow (low) frequency at night be an opening for better contact?

Also from *Awake and Empowered Expo*:

"The use of supplements coupled with a more plant based diet is an excellent way to decalcify the pineal gland. Nascent iodine is an electrified form of iodine that is readily absorbed by the body which helps eliminate fluoride, bromine, chlorine, and other toxins. Not only does nascent iodine remove pineal toxins, but it also stimulates the thyroid, improves mental clarity, and aids in boosting the immune system.

Chaga mushroom is one of the best supplements for pineal decalcification because of its high melanin content. Melanin is produced by the pituitary gland in the presence of melatonin and is a primary pineal food source. Chaga is one of the highest natural sources of antioxidants and helps balance the immune system and prevent cancer, as well as a host of other diseases.

Chaga medicines can be made into powder, tea, and single or dual extracts. Dual extracts are the most potent because they contain both the water and alcohol soluble compounds of the mushroom and are rapidly absorbed into the bloodstream.

Adding magnesium bicarbonate to drinking water not only decalcifies your pineal gland but alkalizes your body and eliminates calcium deposits that have formed around other organs. Its Magnesium in water is absorbed 30% faster than magnesium in food and the bicarbonate amplifies the amount of magnesium absorbed by cells.

The Pineal Gland

Since the pineal is a light sensitive, sun-gazing is a simple way to decalcify it. Sun-gazing is the practice of looking at the sun during the hour of sunset or sunrise in order to absorb its light energy. However, mainstream science has taught us that looking at the sun for even brief periods is harmful to our eyes, so many are skeptical of the process. In addition to pineal decalcification, sun-gazing can improve mental clarity, eliminate toxins, prevent mental and physical disease, slow the aging process, and decrease hunger.

One of the most popular methods of gazing is the HRM method, named after its founder Hira Ratan Manek. HRM recommends that you begin gazing for 10 seconds, increasing 10 seconds each day until you reach a time of 45 minutes. According to this method total, mental clarity occurs at 15 minutes and physical diseases start to disappear around the 30-minute mark. HRM believes after the 45-minute mark is reached, a person can stop gazing and start walking barefoot for 45 minutes per day on bare earth.

This connection with the earth ensures that your "solar battery" remains charged. If this process is continued one experiences feelings of bliss and joy as their intuition supersedes the ego mind. It is recommended that a person only gazes an hour within sunset or sunrise when the UV index is low. When the sun is higher in the sky, the UV rays can be damaging to the retina of the eye.

In order to gain the full benefit, HRM recommends that a person grounds themselves by standing barefoot on dirt or sand while gazing. It is also important to note that you should never gaze while standing on grass because the plant life can absorb sun energy from the body and lessen the benefits. Sun-gazing is a journey for each person and the benefits and experiences are unique based on the intention and needs of the individual.

Western society once thought the third eye was a vestigial organ with no function, but its mysteries are slowly unraveling as science and spirituality merge. The discovery of DMT as a naturally occurring psychedelic adds validity to the complexities and unique nature of mystical experiences.

As people connect with their spiritual nature through pineal decalcification, meditation, and other holistic practices the power of intention, belief, and thought becomes a driving force in the areas of health and spiritual attunement. We have only just begun to tap into the nature of our pineal, but as scientific evidence confirms spiritual knowledge mystical experiences become tangible tools for physical and spiritual growth."

End of second "Awake and Empowered" quote. See: www.awakeandempoweredexpo.com or attend an expo.

Many people receive answers to troubling questions after a good night's sleep. It's my opinion that humans have a way, via the pineal gland, of developing a vibrational frequency whereby better communication with our Maker is not only possible but likely. It seems very important that we all work toward de-calcification of our pineal gland to enhance our God-given attributes, possibly giving us a highly enlightened intuitiveness. Bigfoot does not use fluoride. :)

Chapter 12
Professional Findings

In my first book, *Voices in the Wilderness*, I gave a brief summary of the work Al Berry and I did to establish credibility behind the sounds we recorded in the Sierras of California during the 70's. Dr. Kirlin, quoted below, was neither the first professional who was approached, nor the first to take the recordings seriously.

Syntonic's Research, responsible for working on the Watergate tapes, President Nixon's downfall, was the first to examine our recordings. Their analysis was written about in Al's book, *Bigfoot*, co-authored with Ann Slate in 1976 (See Chapter 19, Al Berry).

> "If Bigfoot is actually proven to exist, the vocalizations on these tapes may well be of great anthropological value, being a unique observance of the Bigfoot in its natural environment."
>
> Dr. R. Lynn Kirlin, Professor of Electrical Engineering, 1978, after a yearlong University of Wyoming-based study of the Bigfoot recordings

Dr. R. Lynn Kirlin

Professional Findings

Dr. Kirlin was however, the first to actually spend a year analyzing and giving the recordings written credibility in a professional book, *Manlike Monsters on Trial*. The book was published after Dr. Kirlin gave his oral presentation to a group of scientists and Bigfoot enthusiasts in 1978.

See: *Manlike Monsters on Trial: Early Records and Modern Evidence*; Halpin & Ames, eds. University of British Columbia Press, Vancouver, 1980.

Dr. Kirlin's Curriculum Vitae:

He held a high-level security clearance until Jan. 2011. Many years of university-industrial contractual projects, algorithm development, performance analysis and testing with software, statistical signal and array processing as applied to 1D, 2D and 3D data, images, speech/hearing, eeg, sonar, HF radar, seismic, communications and DC.

He is extensively published industrial and academic award recipient in Canada and the US. A researcher and consultant with excellent problem-solving and algorithm development skills applicable to statistical detection, estimation and presentation within diverse applications typical of communications sonar, seismology, and radar.

R. Lynn Kirlin received his BSEE and MSEE from U. Wyoming in 1962 and 1963 and his Ph.D./EE from Utah State in 1968. His industrial experience includes communications systems at Martin-Marietta and Boeing 1963-66, computer peripherals at Datel (Wyoming) 1969, and applications software at Floating Point Systems, (Oregon) 1979. He was with the EE Dept. at U. Wyoming from 1969-1986 and has been with the ECE Department at the University of Victoria since 1987.

His major research and consulting associations have included the US and Canadian Naval organizations and oil industries in the areas of statistical seismic and sonar array signal processing. He was an associate editor for the IEEE Transactions on Signal Processing in 1989-91. He is a co-author of the 1998 Best Paper in Geophysics and co-editor and major author of Covariance Matrix Analysis for Seismic Signal Processing.

For Dr. Kirlin's full professional biography, please see: https://www.uvic.ca/engineering/ece/faculty-and-staff/home/faculty/emeritus/kirlinr.-lynn.php

Nancy Logan, Human Sound Expert

"I'm not a tape expert, but I am a human voice sound expert, and…who, or whatever, made these noises has a voice pitch range that is considerably more flexible than that of Homo sapiens sapiens (humans). It goes much lower and much higher." Nancy Logan, 1996.

Narrative of Nancy Logan

"I have been listening to the 'Biggie' tapes again today, trying to pick just one more thing out so that I can give you my best 'statement.' Every time I listen to these tapes, I get chills up and down my spine.

First…I'll give you some of my 'qualifications' to study these sounds…we (my twin sister and me) have what is known as 'perfect pitch,' or 'absolute pitch,' which permits a person to know what a note is without having any reference note played…and to be able to mimic and hear language sounds that are not of that person's 'native language'.

Professional Findings

I've had the experience over and over of being able to reproduce the pronunciation of other languages immediately after hearing them the first time when people who have studied those languages for years can neither reproduce them nor hear them. This talent (has) led me to play lots of musical instruments growing up...at present, I play the flute, the Japanese harp (Koto) and the piano. I have a Russian friend... she thinks I'm a Russian...the bottom line is I have a very sensitive sense of hearing...

I am currently the only interpreter in the State of California who is court certified in Spanish/English and Japanese/English and court registered in French/ English...I also speak Russian and Italian (fluently) although not at the level of an interpreter, which is very fluent. There are only 10 people in the State of California who are currently certified in Japanese/English. There are approximately 600...who are certified in Spanish. That is out of a population of 20 million in the southern California region...about 1% of the people taking these tests pass. Besides the languages I speak, I also (have) studied others, such as Chinese and Arabic...

I have listened to these purported Sasquatch tapes quite a few times. I listened to the one with Jonathan Frakes again today, too, as well as the other one you sent me...it seems that every time I listen to them, I get something new. Anyway, here is what I now have to say about them.

First of all, I don't see how they could be fake. For one thing, the creatures' articulation is incredibly fast at certain points, but in order to make the voices sound as low in places as they do, a tape of a person speaking would have to be slowed down. That's basically impossible with the speed of articulation being what it is.

I'm not a tape expert, but I am a human voice sound expert, and...who, or whatever, made these noises has a voice pitch range that is considerably more flexible than that of Homo sapiens sapiens (humans). It goes much lower and much higher. I don't think that any Homo sapiens sapiens will ever be able to truly mimic those sounds, not even someone who's good at it like me...

"Secondly...the sounds on the tapes sound like they are apes (BIG apes) sometimes and they are humans (BIG humans) sometimes. In places there are definite vocalizations that sound like some type of semi-primitive language...there are certain places on the tape where they seem to be repeating 'words' or sounds, and they are making these noises at an incredible rate of speed and with a natural-sounding intonation. They are not just saying 'ga-ga gu-gu ba-ba, etc.' which is what hoaxers would more than likely do...

"I believe that some sort of primitive communication is going on in the form of primitive language. The first time I listened to the tapes, I thought it was linguistically a little more sophisticated than I do now. After listening to them again, I think that the creatures are a little more animal sounding, but I still think it is language. I challenge anyone to make those exact same noises with the exact same pronunciation and intonation at that speed...

"Some of the sounds on the tape clearly sound like some sort of posturing, testing behavior. Some sound clearly like they are trying to communicate with you. The passage on the tape where you keep saying 'they are in the rocks' and you are trying to imitate them is really interesting. After listening to it one time, it sounded very much to me like people making noises, but after listening to it a few times, I could faintly hear that they are making similar noises to those that they make when they are closer, such as those kind of puffing noises, and the ooh-hoo ooh-hoo noises.

The chattering. This is really exciting to me, because it is just exactly what you would expect from language, and it indicates a certain kind of linguistic pattern to me. Also, there are at least two creatures in that sequence, and probably more.

Without some sort of context, it's almost impossible to tell what they might be saying. On one spot of the tape, an airplane goes by and they seem to get very excited and not very happy about it. Maybe those are Sasquatch swear words. They also seem to have quite a bit of vocal interaction among themselves, which is more evidence of language...

Anyway, here is what I believe:

1. I don't think Homo sapiens sapiens can make all of these noises and in this fashion. I've heard and practiced a lot of known human language sounds and vocalizations and these don't fit any of them and I can't make these noises. One example of this is that the 'whistling' noises they make sound to me like they are coming from the throat. I don't know any human who can do that, though with practice someone might???

[Sound analysis of the whistles shows they are harmonic and do not originate at the lips, as would an ordinary human whistle AB]. This (that the noise is not human) does not rule out that it is language. The vocalizations seem to have some elements of language to me, i.e. certain repeated phoneme patterns and a certain organization to the chattering.

2. There seems to be some sort of objective to what they are doing, maybe from curiosity. It is not just a situation where animals are circling your camp and hooting and whooping.

3. These creatures have an incredible range of voice and ability to make different noises. Possibly able to imitate the noises of many forest creatures?? Maybe a lot of people have heard them throughout the years and just didn't know it because they thought they are animals.

4. They are trying to communicate with you. If they are anything like us, the chimps and the gorillas, then they probably have a healthy dose of curiosity in their psychological makeup...

5. It would take an incredible amount of training for a human to make these noises so fluently and spontaneously. The noises also include vocalizations made with what sounds like parts of their vocal tract that native English speakers would have tremendous trouble in learning.

6. I mentioned (previously) that what sounded like the 'female' voice on the tape is much more vocal than the male, and that this mimics human society. The 'female's' loquaciousness is interesting. If someone had gone to the trouble to go into the woods on purpose to fake tapes, I'm certain that they would have made it sound like the 'male' was the one doing all the talking, just as your tape assumes that the 'little' one that can be heard talking is a boy. Maybe it's a girl.

I don't think these tapes are fake." Nancy Logan

R. Scott Nelson, Crypto-Linguist

"These creatures have a 'Complex Language', by the human definition of language."
Scott Nelson, Crypto-Linguist

Mr. Nelson is retired from the U.S. Navy as a Crypto-Linguist with over 30 years' experience in Foreign Language and Linguistics, including the collection, transcription, analysis and reporting of voice communications.

He is a two-time graduate of the U.S. Navy Cryptologic Voice Transcription School (Russian and Spanish) and has logged thousands of hours of voice transcription in his target languages as well as in Persian. He is currently teaching Russian, Spanish, Persian, Philosophy and Comparative Religions at Wentworth College in Missouri.

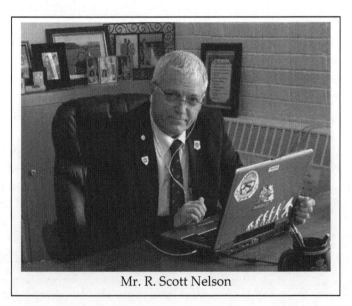

Mr. R. Scott Nelson

Professional Findings

Vetting Mr. Nelson:

(From private correspondence concerning Mr. Scott Nelson)

By: **J. Edward Boring, Chief Knowledge Officer**
Defense Language Institute Foreign Language Center
1759 Lewis Road, Bldg 614, Suite 251
Presidio of Monterey, CA 93944.

Speaking in general, especially with regard to what the subject may have learned while a student at DLIFLC, odds are strongly in favor of him having learned more than enough about language origins, structure, formation, and development to be well-qualified to complete the research and documentation to which you refer.

As near as I can tell, from your statements, the subject has not translated the recordings. He has simply analyzed them and applied phonetic structure to the sounds on the recordings. This is well within the skill levels expected of any graduate of DLIFLC. If the subject's skills have been further enhanced by years of study and professional experience, I would expect an even more scholarly capability to listen, transcribe, analyze, document, organize, and explain (phonetically at least) any "unknown language."

In fact, the entrance exam that qualifies (or not) a person for enrollment at DLIFLC requires the examinee to learn a "new language" on the spot, while taking the examination, correctly answering questions presented in the "new language" (which is not a known extant human language), thus exhibiting the examinee's aptitude for linguistic studies.

Hence, I have no difficulty at all stating that IF the subject is, in fact, a "two-time graduate of DLI" he would be utterly qualified to do what you say he has done, and more besides. At the end of the day, this type of research and analysis are fundamental to a career cryptolinguist's profession.

Professional Findings

Interview with Mr. Nelson June 2010, www.nabigfootsearch.com

Question: Mr. Nelson, Could you please explain in layman terms why you think language exists in the recordings you've heard?

Scott:

It might be best to start here with my arguments for the three conclusions that I drew almost immediately upon hearing the Sierra Sounds for the first time. Probably the best way to do that is to copy some of the notes I use when I present my study at conferences and symposiums:

After that first quick review of the samplings of the Sierra Sounds, there were three facts that were immediately evident to me:

- The vocalizations are not human (as we currently define human);
- The creatures were speaking in a complex language (by the human definition of language); and
- The tapes could not have been faked.

First: The voices are not human. The creatures on the Berry-Morehead Tapes are producing sounds that humans cannot make. Their vocal range is far too great; much lower and much higher than humans are capable of producing. This fact is corroborated by the Kirlin study.

Additionally, the volume and resonance of many of the vocalizations they produce is far beyond the ability of humans. However, the most striking element to note is the prosody of utterance, or the tempo at which each utterance is delivered, as well as the speed at which the conversational turns take place, with the creatures almost stepping on each other in their discourse. For the majority of the utterances, the rate of deliverance is at least twice that of humans.

My second conclusion: It is a complex, human-like language. What did I recognize in the vocalizations that told me that it was language? First are the articulated phonemes (individual units of phonetic sound) so similar to our own that it must be assumed they are produced by the same apparatus that we possess, namely, the tongue, the teeth and the lips, along with the entire tracheal tree, oral cavity and nasal cavity.

I have isolated 39 different phonemes, all common to human language. Phonemes combine to form morphemes, or individual units of meaning which we commonly call syllables or minimal words. These are evident throughout the tapes, repeated in conversational turns and morpheme streams characteristic only of language.

We find discourse (conversational turns of utterance); query inflection and direct response; imperative or persuasive inflections; expression of emotion, intimidation, negation and even ritual. These vocalizations exhibit characteristics that are conventional, automatized, arbitrary and creative; all of which are properties of human language.

In brief, there are so many characteristics of human language evident in the tapes that we must assume that even those elements that cannot yet be known, such as grammatical categories, are also present in this language.

Finally to my third conclusion: The tapes could not have been faked. While serving as a crypto-linguist working with Naval Intelligence, I trained in every form of deceptive voice communication imaginable, including slowing the tape; speeding it up; modulation of tone and pitch; playing tape backward and distortion of every kind. None of these techniques is evident here.

I was a Russian analyst so I trained in all of the Soviet tactics of deception. They are the best in the world at deceptive communication techniques, but even their best effort could not have produced these vocalizations; and certainly, no one could have done it in 1974. What initially led me to conclude that the tapes were not fake is that in numerous instances the humans and the creatures are speaking at the same time; vocally stepping on each other. This cannot be done without leaving trace evidence (also confirmed by the Kirlin Study).

At this juncture, to claim that these vocalizations were faked, one would have to argue that a secret cabal, comprised of several ingenious conspirators, was so determined to deceive the world, that they invented their own language, modulated their vocalizations to frequencies above and below the ability of humans, harassed a small group of well-armed hunters, over a period of several weeks in successive years and threw in numerous cognate words and expressions to boot. It is now more reasonable to defend the existence of an undocumented creature than it is to believe in such a conspiracy.

The Ketchum Dichotomy

In 2013 Dr. Melba Ketchum, Ph.D. geneticist, and practicing veterinarian, released the DNA results of a study which she conducted on purported Bigfoot samples. The mtDNA (maternal side) was found to be human. Therefore, that's why some scientists, who had reviewed her findings say, "The samples were shown to be human — done," or so they say!

However, the nuDNA, with information coming from two parents, one male and one female, rather than matrilineally, as in mitochondrial DNA, was shown to not be in any Genbank DNA sequence. No match was found; an unbelievable fact considering the Genbank band 224 billion bases in their catalog. See http://sasquatchgenomeproject.org/

Professional Findings

Dr. Ketchum's results created a firestorm in the science community. She was attacked, and her results were dismissed on several fronts. In brief the objections were:

- It was argued Dr. Ketchum's samples were contaminated through careless handling in the gathering process, or in the laboratory used to examine the samples.
- Dr. Ketchum was privately funded by a wealthy Bigfoot enthusiast and not an institution or a government agency.
- Dr. Ketchum published results in a journal primarily controlled by her.

Through these arguments, the peer review factor was aggressively dismissed. There are also philosophical reasons Dr. Ketchum, thus far, may have been denied the privilege of claiming one of the most groundbreaking discoveries of the century. If human hybrids are established to be on our planet it will affect history, sociology, biology, and other core beliefs long held by theological and scientific communities alike. Economic implications could also be strong and atrocious to some profit based sectors.

June 1992: The author and Al Berry on a precipice in British Columbia.

Excerpt from the following chapter:

"For a person traveling at 99% the speed of light, time slows for them by a factor of 7. If they were to travel to a star 7 light years away, at 99% of the speed of light, it would take them 1 year; but to an observer on Earth, it would have seemed like 7 years.

However if that person attained 99.999% the speed of light, only 1 year would pass onboard for every 223 years back on earth."

www.astronomytrek.com

Chapter 13
Quantum Time

> "Time travel used to be thought of as just science fiction, but Einstein's general theory of relativity allows for the possibility that we could warp space-time so much that you could go off in a rocket and return before you set out."
> Stephen Hawking

You may remember, a few years back a Bigfoot was reportedly shot in California — the guy said that he didn't miss, but couldn't locate the body. Fred Beck also claims to have shot one in Oregon during an encounter in 1924. Beck said he didn't miss, but there was never a body to be found in either instance. Over the years I've heard several stories of hunters shooting a Bigfoot, but again, a body was never recovered. How can this be? Is it because of our inept ability to shoot straight, track better, or could there be another more interesting explanation through quantum physics?

After reading the following time-travel article you might begin to see how science could support the possibility of a disappearing 8ft body. According to quantum physics; time, as we know it, only exists in our three-dimensions. This may sound a little outlandish, but let's read on.

Scientists from the University of Queensland, Australia, have used single particles of light (photons) to simulate quantum particles traveling through time. They showed that one photon can pass through a wormhole and then interact with its older self. Their findings were published in Nature Communications. By: Vandita in 2015, www.anonews.co/aus-time-travel/

"The source of this time travel conundrum comes from what are called "closed timelike curves" (CTC). CTCs are used to simulate extremely powerful gravitational fields, like the ones produced by a spinning black hole, and could, theoretically (based on Einstein's theory of general relativity), warp the fabric of existence so that spacetime bends back on itself – thus creating a CTC, almost like a path that could be used to travel back in time.

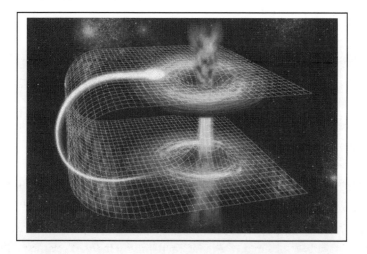

According to "Scientific American," many physicists find CTCs "abhorrent, because any macroscopic object traveling through one would inevitably create paradoxes where cause and effect break down." However, others disagree with this assessment.

In 1991, physicist David Deutsch showed that these paradoxes (created by CTCs) could be avoided at the quantum scale because of the weird behavior of these fundamental particles that make up what we call matter. It's well known that at the quantum scale, these particles do not follow the rules that govern classical mechanics, but behave in strange and unexpected ways that really shouldn't even be possible. Source: collective-evolution.com

"We choose to examine a phenomenon which is impossible, absolutely impossible, to explain in any classical way, and which has in it the heart of quantum mechanics. In reality, it contains the only mystery."

Richard Feynman (1918-1988) American physicist and a Nobel laureate of the twentieth century: From Radin, Dean. *Entangled Minds: Extrasensory Experiences In A Quantum Reality.* New York, Paraview Pocket Books, 2006.

"In the quantum world, paradoxes that we don't understand are common findings, but this should not deter people from taking this science seriously. Even Einstein didn't believe a lot of quantum theory, but I'd like to think that if he were alive today, he would definitely be having some fun, given all of the recent breakthroughs.

It's intriguing that you've got general relativity predicting these paradoxes, but then you consider them in quantum mechanical terms and the paradoxes go away." Tim Ralph, University of Queensland physicist

The Experiment

Tim Ralph (quoted on pg 157) and his Ph.D. student Martin Ringbauer simulated a Deutsch's model of CTCs, according to Scientific American, "testing and confirming many aspects of the two-decades-old theory." Although it's just a mathematical simulation, the researchers (and their team/colleagues) emphasize that their model is mathematically equivalent to a single photon traveling through a CTC.

Nothing has actually been sent back through time, though; to do that, scientists would have to find a real CTC, which has yet to happen as far as we know. Of course, there always remains the possibility that black budget science has.

Think in terms of the 'grandfather paradox,' a hypothetical scenario where someone uses a CTC to travel back through time to cause harm to their grandfather, thus preventing their later birth. Now imagine a particle going back in time to flip a switch on the particle-generating machine that created it – this is a possibility that these physicists say they have shown through their simulation. Source: Scientificamerica.com.

You can read the specifics of the experiment at https://www.scientificamerican.com/article/time-travel-simulation-resolves-grandfather-paradox/.

Quantum Time

Why this is a High Probability

In this author's opinion, there is no doubt time travel is possible. Why do I believe this? Well, it's because we know one hundred percent that superposition is real on a quantum scale.

"The maddening part of that problem is that the ability of particles to exist in two places at once is not a mere theoretical abstraction. It is a very real aspect of how the subatomic world works, and it has been experimentally confirmed many times over."

From an article by Tim Folger in Discover Magazine, June 2005. Where bold per original.)

"One of the supreme mysteries of nature... is the ability, according to the quantum mechanic laws that govern subatomic affairs, of a particle like an electron to exist in a murky state of possibility — to be anywhere, everywhere or nowhere at all — until clicked into substantiality by a laboratory detector or an eyeball." By: Dennis Overbye, New York Times, March 12, 2002

This means that one particle can exist in multiple states at one time. This is best demonstrated by the quantum double slit experiment. Recent experiments have also confirmed quantum entanglement, showing that space is really just a construct that gives the illusion of separation. One thing that suggests there is a high probably of time travel, in conjunction with the experiment mentioned in this article, is the fact that there are experiments showing that particles can actually be entangled through time.

This is illustrated by what is called the "Delayed Choice Experiment."

Like the quantum double slit experiment, the delayed choice/quantum eraser has been demonstrated and repeated time and time again. For example, physicists at The Australian National University (ANU) have successfully conducted John Wheeler's delayed-choice thought experiment. Their findings were published in the journal "Nature Physics."

In 2007 scientists in France shot photons into an apparatus and showed that their actions could retroactively change something which had already happened. This particular experiment illustrates how what happens in the present can change what happened in the past. It also shows how time can go backward, how cause and effect can be reversed, and how the future caused the past. See Collective-Evolution.com

"If we attempt to attribute an objective meaning to the quantum state of a single system, curious paradoxes appear: quantum effects mimic not only instantaneous action-at-a-distance but also the influence of future actions on past events, even after these events have been irrevocably recorded." – Asher Peres, a pioneer in quantum information theory.

Although we do not have access to a CTC quite yet, there are good reasons to believe that this type of time travel is possible at the quantum mechanical level, and that is why I chose to mention these other experiments, to show that 'time' doesn't even really exist as we think it does.

Why these same quantum mechanical laws have not been observed on the macroscopic level is yet to be understood, but physicists are working on the problem. For example, in 2012 physicists David Wineland and Serge Haroche received the Nobel Prize in physics for demonstrating how "quantum weirdness" could not only exist at the subatomic micro-world level but also show itself in the macro-world. At one time, superposition was only thought to exist in the inaccessible quantum world, but not anymore. We know it's possible, we just haven't figured out how. We do, however, seem to be getting closer to finding out."

From "Science News," Oct 2012, Nobel Prize 2012

Perhaps one day, we will have determined the key to this puzzle and be able to observe large objects like cars, humans, apples, and oranges behave in the ways that matter does on a subatomic level, and perhaps one day we will find a wormhole, or a CTC in space, to conduct actual experiments that go beyond theory. That being said, a lot of what used to be considered theoretical in quantum physics is no longer theoretical, like quantum entanglement.

Interesting Facts About Time, The Fourth Dimension, And Time Travel. www.Astronomy Trek:

"Time is perhaps the greatest mystery of all and is deeply wrapped up in our conscious experience of things.

For Newton time was absolute, with Einstein time became more flexible and relative in scope. However, no one has been able to fully explain what it really is.

1: Time Is The Fourth Dimension

Simply stated, the first three dimensions are used to specify an object's location/movement in space (forward-backwards, left-right and up-down), while the fourth dimension locates its position in time. All four dimensions are used to specify completely the location or dynamism of an object in space. Collectively the four dimensions are inseparably interlinked and known as space-time.

2: Three-Dimensional Creatures

Being three-dimensional creatures (possessing length, width, and height), humans are unable to see the fourth dimension as our physical world is constructed within these three physical dimensions. We might feel or perceive time's presence, but we can never actually detect it with our three-dimensional senses because it extends beyond our universe. Humans only perceive the fourth dimension time as memories lodged at variable intervals, the result of which is our apparent perception of time moving forward in a straight line.

Time still exists as a dimension and objects can cross it in a similar way as they do the others, although three-dimensional like humans can only move in one direction forward through time. If we could see an object's fourth-dimensional space-time (or world-line) it may resemble a spaghetti-like line stretching from the past into the future showing the spatial location of the object at every instant in time.

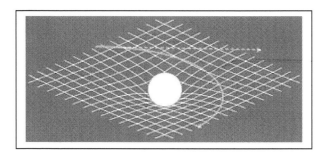

3: Space and Time Inseparable

Space And Time are simultaneous phenomena (like mass and energy), and together form the fabric of the universe known as space-time. A demonstration of four-dimensional space-time's inseparability is the fact that, as astronomers often remind us, we cannot look into space without looking back into time. We see the Moon as it was 1.2 seconds ago and the Sun as it was 8 minutes ago.

Also, in accordance with Einstein's General Theory of Relativity, a massive object in space stretches the fabric of both the space and time around it. For example, our Sun's mass bends its surrounding space so that the Earth moves in a straight line but also circles within the Sun's curvature in space. The Sun's effect on time is to slow it down, so time runs slower for those objects close to the massive object. Interestingly, gravity is the result of mass stretching the fabric of the space-time around it. Gravity also has an infinite range such that no matter how far apart two masses are in space they will always experience some gravitational pull towards each other. Theoretical physicists have tried to explain this phenomenon in terms of gravitons, S-Theory, and M-Theory, but even today a successful quantum theory of gravity is yet to be found.

4: Time and The Speed Of Light

A property of light is that it always travels at the same constant speed in a vacuum of 186,000 miles a second (700 million mph) and you can't go any faster. The reason for this is that mass increases with speed all the way to infinity and so an infinite amount of energy would then be needed to travel beyond the speed of light.

The equation states that: speed = distance ÷ time; therefore if the speed of light (c) is to remain fixed then the distance and time in the equation will need to change. What actually happens is that time and distance are 'relative' to one another, and as you travel close to the speed of light, distances become shortened while time is lengthened. This is explained in Einstein's Theory of Special Relativity.

For a person traveling at 99% the speed of light, Time slows for them by a factor of 7. If they were to travel to a star 7 light years away, at 99% speed of light, it would take them 1 year, but to an observer on Earth, it would have seemed like 7 years. However, if that person attained 99.999% the speed of light, only 1 year would pass on-board for every 223 years back on Earth. Finally, you don't need to travel at light speed for time dilation to occur but you won't notice the effects until you go extremely fast." End of quote.

"When we see science fiction movies, space and time warps are common. They are used for rapid journeys around the galaxy, or for travel through time. But today's science fiction is often tomorrow's science fact." Dr. Stephen Hawking.

What does time, as explained by quantum physics, have to do with an absent Bigfoot carcass? At the beginning of this chapter, I mentioned that hunters claimed to have shot a Bigfoot, but the body was never found. According to many physicists time travel is not just possible, but a mathematical fact — perhaps Bigfoot literally dodged the bullet!

Chapter 14
Two Minds

"Ego is the immediate dictate of human consciousness:"
Max Planck

Diminishing the ego may get you closer to Bigfoot. But, do you want to get closer to Bigfoot — that is the question. I obviously don't know if any of them have what we call an "ego." However, as humans, I believe if we allow our egos to get in front of our mission of discovery, our efforts are compromised. Therefore, I thought this chapter might help in that regard — so I wrote it.

Often we've heard the phrase, "The devil made me do it." Or, "If there wasn't any bad luck, I'd have no luck at all." Could we be attempting to get the monkey off our backs by trying to 'not' take responsibility for our own actions? Could your ego be getting in the way?

In psychological terms, the ego is the part of the psyche that experiences the outside world and reacts to it, coming between the primitive drives of the 'real you' and the demands of the social environment, represented by the ego. Source: Dictionary.com

It feels good to know that someone thinks we're wonderful and it's extremely difficult for most of us not to hang on to a compliment. So just how much of the ego must die to become who we're supposed to be? The Apostle Paul said, "I die daily." (1 Corinthians 15:31).

"No one can serve two masters. Either you will hate the one and love the other, or you will be devoted to the one and despise the other. You cannot serve both." (Mathew 6:24).

So now I ask, do we serve the ego master or the One master who's energy resides in ALL humans?

We are all designed to be part of a 'Oneness', an energy coupled to a universal collectiveness. And I don't think that Oneness is in woo-woo land. I have always believed that how and what you eat can determine your health. So too, what you feed your brain will determine how you create your own paradigm.

> "There is a science called neuroplasticity. It means that our thoughts can change the structure and function of our brains. The idea was first introduced by William James in 1890, but neuroplasticity was soundly rejected by scientists who uniformly believed the brain is rigidly mapped out, with certain parts of the brain controlling certain functions. If that part is dead or damaged, the function is altered or lost. Well, it appears they were wrong.
>
> Neuroplasticity now enjoys wide acceptance as scientists are proving the brain is endlessly adaptable and dynamic. It has the power to change its own structure, even for those with the severe neurological afflictions. People with problems like strokes, cerebral palsy, and mental illness can train other areas of their brains through repetitive mental and physical activities. It is completely life-altering.
>
> Many fear to do something new because we don't want to fail, or we were told it would be impossible. The truth is, we can do most anything if we take action, stop negative thinking, and shift our perceptions of the truth about our abilities.

Two Minds

It was believed until recently that the human brain, which consists of around 100 billion neural cells, could not generate new ones (the generation of new neurons is also known as neurogenesis). The old model assumed that each of us was born with a finite number of neural cells and when a cell died no new cell could grow.

This old model of the brain's inability to regenerate new nerve cells is no longer relevant. It has been proven that certain areas in the brain can generate fresh cells. This new understanding of neural cell generation is an incredible discovery. Another misconception was that the brain had an inability to create new neural pathways. It was once believed that the human brain had a relatively small window to develop new pathways in our life span, then after that, the pathways became immutable.

This old theory thought our ability to generate new pathways dropped off sharply around the age of 20 and then became permanently fixed around the age of 40. New studies have shown through MRI brain scanning technology, that new neural cells are generated throughout life as well as new neural pathways. Even the elderly are capable of creating measurable changes in brain organization. These changes are not always easy but can happen through a concerted focus on a defect area." Source: Norman Doidge M.D., brainhq.com

What does any of this ego and brain stuff have to do with Bigfoot? I think if we can get our ego out of the way these Bigfoot beings may come out and interact with us so we can better understand them. If we're thinking we need that 'Money Shot' with the camera it's probably our ego working and that might be sensed by these beings. I think Bigfoot is in tune with nature (Oneness) and are able to sense our exploitations.

So I say, "get the ego out of the way." Most often it's what stands there, licking its chops. This allows us to show the positive energetic 'being' that is inside — serve 'that' energy. Be that person who you were designed to be — the embodiment of God. The One and only master we are to serve — not our ego. And it will serve us to remember that "the brain should be subservient to the heart."

> ....they are darkened in their understanding andseparated from the life of God because of the ignorance that is in them due to the hardening of their hearts." Ephesians 4: 18

> "A good man out of the good treasure of his heart bringeth forth that which is good, and an evil man out of the evil treasure of his heart bringeth forth that which is evil: for of the abundance of the heart his mouth speaketh." Luke 6:45

The following is from an informative biological article about how the heart and brain are connected: David Math, *TruSparta*:

Two Minds

"The strongest aspect of the electromagnetic field within each and every single one of us is the Heart center. The human Heart is now documented as the strongest generator of both electrical and magnetic fields in the body. Important, because we've always been taught that the brain is where all of the action is. While the brain does have an electrical & a magnetic field, they are both relatively weak compared to the Heart.

The Heart is about 100,000 times stronger electrically & up to 5,000 times stronger magnetically than the brain. Important, because the physical world – as we know it – is made of those 2 fields: electrical & magnetic fields of Energy. Physics now tells us that if we can change either the magnetic field or the electrical field of the atom, we literally change that atom and its elements within our body and this world. The human Heart is designed to do BOTH."

The more I read articles like this and compare those to ancient texts, the more assured I am about my outlook…especially as I read the Red-Letter edition of the New Testament. Studying just the words of Christ and getting to the core meaning is very eye-opening.

Does 'Cold' exist, or is it the absence of Heat? Does 'Darkness' exist, or is it the absence of Light? Evil is the absence of Love. If someone is not connected with the oneness of God's energy, that "Someone" is living life on their own terms, using the outside senses, working from the ego and serving themselves. We can only serve one master. So, it's worth mentioning again, this time from the book of Luke, and this time we have "Mammon," meaning money.

> "No man can serve two masters: for either he will hate the one, and love the other; or else he will hold to the one, and despise the other. Ye cannot serve God and Mammon." Luke 16:13

From Wikipedia: "In the Bible, mammon is commonly thought to mean money, material wealth, or any entity that promises wealth, and is associated with the greedy pursuit of gain."

Chapter 15
Invisibility

"In a time not distant, it will be possible to flash any image formed in thought on a screen and render it visible at any place desired. The perfection of this means of reading thought will create a revolution for the better in all our social relations." Nikola Tesla

Why does it surprise us when we hear that something can become invisible? Maybe it's because so much of our understanding of this world comes to us through our eyes. Sometimes if we say, "I see", it usually means you understand. In this chapter, we will discuss two types of invisibility, i.e., classical and quantum.

Cloaking Demonstration

Invisibility Cloak Technology

"Professor John Pendry (1943): Physicist and winner of the prestigious Newton Medal (the highest honor of the UK's Institute of Physics). Pendry first published an idea for an "invisibility cloak" in 2006. In 2013 he discussed invisibility cloak technology, at the Imperial College in London, e.g., Metamaterials and the Science of Invisibility. He is a professor of theoretical solid state physics there. This is a fairly new technology in the works.

As we know, light can bend when going through a sphere, even through water. However, Dr. Pendry's idea is more subtle — going through a material where light is bent gradually around an object that is covered with the Metamaterial that he speaks of. In order to make an object no longer visible, it is necessary to bend or channel the light around it in such a way that what the viewer sees is what stands behind the object rather than the object itself; this gives the impression of being able to see through it.

In July 2016, scientists at Queen Mary University of London (QMUL) have made an object disappear by using a material with nano-size particles that can enhance specific properties on the object's surface. For the first time, researchers demonstrated a practical cloaking device "that allows curved surfaces to appear flat to electromagnetic waves," according to the university. The researchers used a nanocomposite medium to coat a curved surface about the size of a tennis ball. Project co-author Professor Yang Hao said, "The design is based on transformation optics, a concept behind the idea of the invisibility cloak." Queen Mary University of London Media News 2016.

"Researchers have been able to quantify fundamental physical limitations on the performance of

cloaking devices, a technology that allows objects to become invisible or undetectable to electromagnetic waves including radio waves, microwaves, infrared and visible light. The researchers' theory confirms that it is possible to use cloaks to perfectly hide an object for a specific wavelength, but hiding an object from an illumination containing different wavelengths becomes more challenging as the size of the object increases."
Source: Crystallinks.com

Author's note:

Many of these studies fall under the realm of Classical Science. But for years I have heard people refer to something a bit different — they saw a Bigfoot disappear, actually, cloak in front of their eyes. For years I discounted those accounts — until I began to look into quantum physics and began to understand (to some degree) how that could actually happen. It's all about frequency, electrons, and waves.

There is something called Human Involuntary Spontaneous Invisibility, in which someone disappears without explanation. Unlike classical science, it's about shifting vibrational frequency.

By: Donna Higbee: www.crystalinks.com:

Human invisibility has been written about for centuries. Indo-European and pre-Aryan shamanistic beliefs accompanied the peoples who eventually migrated into the Indus Valley (approx. 2,500-1,500 BCE). Here, men and women of great spiritual attainment, superior knowledge, and extraordinary powers came to be called rishis.

Invisibility

The Vedas, which form the basis of Hinduism, emanated from the teachings of the rishis beginning about 1,000 BCE. In these texts, we find descriptions of the rituals and techniques of the Hindu priests, sounding very much like the magical and shamanistic abilities of sorcerers, magicians, and shamans.

Later in Hinduism, around 700-300 BCE, we find secret doctrines, called the Upanishads, which were written for students. Within the Upanishads, there is a section called the Yogatattva, which gives the rich mystical philosophy of the discipline and theory of practice for attaining knowledge of the essence of God.

A serious student of raja yoga was taught that certain supernormal powers, called siddhis, were a natural outcome of gaining mastery over one's mind and environment, and were used as valuable indications of the student's spiritual progress. One of these yogic siddhis was human invisibility.

Patanjali, the author of the Yoga-sutra, which is one of the earliest treatises among the early Indian writings, attempts to describe the process whereby human invisibility occurred. He says that concentration and meditation can make the body imperceptible to other men, and "a direct contact with the light of the eyes no longer existing, the body disappears."

The light engendered in the eye of the observer no longer comes into contact with the body that has become invisible, and the observer sees nothing at all. There is not a lot written about how this occurs; the explanation of the process whereby invisibility was brought into being was most likely left up to the

Invisibility

teacher to impart to the student directly.

From the thirteenth century on, numerous texts in Europe refer to similar abilities, performed by sorcerers and magicians who had the power to make themselves invisible, like the shamans (both ancient and modern), and the yoga masters in India. Some other cultures in which shamanism (and the ability to vanish) has played a major role are the Aborigines of Australia, the archaic peoples of North and South America, and the peoples in the Polar Regions.

Rosicrucianism began in Europe in the fifteenth century. Among the papers of that time, there are a number of them that talk about invisibility. A brother in the Rosicrucian fraternity wrote a paper on how to walk invisible among men, and there is evidence that this was being taught in those early days.

H. Spencer Lewis, the founder of the Ancient and Mystical Order Rosae Crucis in San Jose, California, stated that one can gain invisibility with the use of clouds. He says that clouds or bodies of mist can be called out of the invisible to surround a person and thus shut him out of the sight of others.

According to Lewis, this secret practice is still taught in the mystical schools of today. The written literature on this subject supports the statement that the cloud is the basis of the Rosicrucian invisibility secret.

John Macky was an early Masonic leader (the early Masons were believed to be an offshoot of the Rosicrucians). He taught a method whereby any man could render himself invisible.

Invisibility

Another offshoot of the Rosicrucian fraternity, the 'Hermetic Order of the Golden Dawn,' left manuscripts describing the Ritual of Invisibility. These manuscripts talk about surrounding yourself with a shroud, which is described as looking like "a cloud."

It is said that Madame Blavatsky, of the 'Theosophical Society,' witnessed this invisibility for herself and was actually given the secret, thereafter accomplishing this for herself on several occasions in front of witnesses. The literature on the spiritualists in the U.S. shows that there is no doubt they, too, knew about the cloud and its creation.

What is this cloud? We are looking for something that is between empty space and actual physical matter, something unseen by the naked eye but very much in existence. The Rosicrucian manual tells us that the first form into which spirit essence concentrates preparatory to material manifestation is electrons. When spiritual essence gathers into very minute focal points of electrical charge (due to certain conditions), we have the creation of electrons.

Science reports that such a cloud of free electrons will absorb all light entering it; it will not reflect nor refract light waves, nor are light waves able to pass through a human being. Consequently, the observer's eye sees nothing there and the person surrounded by such a cloud is invisible. Since light is necessary for human sight, when there are no reflected or refracted light waves bouncing off a person and hitting the observer's retina, the person is not able to be seen and is not visible under normal circumstances.

How is this cloud created intentionally? That is difficult to say. There are references to and descriptions of invisibility and its creation in the writings of secret societies, but most people don't have access to these writings. One could go to India and become an apprentice or student of an Indian guru or teacher to learn these techniques, but that probably is not practical in modern life. To the everyday person, the knowledge of how invisibility works is a mystery.

With spontaneous involuntary invisibility, it is possible that people are forming a light-absorbing, free-electron cloud around themselves. They are doing it unknowingly and without knowledge of the method. Perhaps they are able to master this skill in another reality which is affecting them here. Since some kind of focused mental thought process must be employed to make the cloud form around oneself, then it might be that these people are doing this unconsciously. Telekinesis is often done unconsciously.

People having these invisibility experiences seem to have higher than average psychic abilities. Possibly they are able to traverse other dimensions and command natural forces, knowingly or unknowingly." End of quote

Invisibility

Becoming Invisible

"With practice, you can become invisible by learning to spiral your personal grid to a higher frequency.

- Begin by sitting very still, relaxing, taking a long deep breath.
- Focus your energies on the electromagnetic energy field, the light field, auric field, that surrounds you.
- Envision the particles in that light field speeding — spiraling upward so they no longer reflect your light.
- Take your time.

If you are successful, you will disappear, the lasting effect varying. Of course, if this works and if you're by yourself, you may have difficulty in proving that it happened to anyone." By Erik Dege: http://invisibility100.tripod.com

Invisibility in Native American Legend

There are many Native American and First People legends depicting invisibility. The following legend was selected because it describes invisibility by Bigfoot. Source: *The Oregon Bigfoot Highway*, Willamette City Press, LLC, 2015; by permission.

"The Clackamas Indian legend states that for a Bigfoot to become accepted into Bigfoot society as an adult, as a warrior, or, as a mature individual, the Bigfoot must present itself in full frontal view of humans three (3) times and not be seen.

The implications of this legend are enormous. Some are obvious. For instance, to prove the three unseen experiences, other Bigfoots probably were present to watch the events and were also not seen.

Invisibility

Another implication is how do they do it? Can they control human minds? Is there a molecular transformation component?

These are important present-day questions. By selling their cloaking techniques to the military Our Bigfoot Friends could become billionaires and buy their own gated forests.

But seriously, and most importantly, to establish a basis of the legend, the Clackamas Indians must have repeatedly seen Bigfoot in various attempts to gain mature status. Perhaps they saw Bigfoot young ones, or Bigfoot adults with impaired mental development who were not capable of making themselves invisible; or possibly adults who exposed themselves not knowing a human was watching.

Another likelihood is the Native Americans encountered advanced adolescent Bigfoot attempting invisibility skills, but only saw parts of the creature. Such an event might be very disturbing indeed. But however appearances materialized, the Clackamas Indians believed the creatures were real. And since there is no legend about conflict, we assume the tribe lived in peace and harmony alongside the Bigfoot." End of quote.

So, does Bigfoot know of an invisibility technique that we have yet to fully understand? Do they know how to change their vibrational frequency to become unperceivable to our eyes?

Cusco, Peru, 2014
Cusco, in the Peruvian Andes, was once capital of the Inca Empire. Note the puzzle-like shape of the huge boulders at the base—placed by a pre-Incan culture. The Incas built on top of that base, and then the Spaniards.

Chapter 16
Is There a UFO/Alien Connection with Bigfoot?

Years ago when I first began interviewing people regarding a sighting, many would also tell me about strange lights, or an actual UFO that was associated with their Bigfoot sighting. Although we had strange lights and unusual happenings around our Sierra Camp, I just didn't want to be pulled into the paranormal circle — as I've already stated, I don't like that word. When academics, the straight people, first hear it they want to put you in a strait jacket.

As strange as it sounds, during the summer of 2016, while at the camp, Keri and I were watching stars early one evening when a two-foot elongated, bright light came floating by us. It had a slightly golden glow, but wasn't blinding us. It was controlled, not moving fast, and we watched it for several seconds, going between the trees about forty yards away — that was truly exciting.

I didn't know what to think; was it going to come closer? It stopped by our old toilet area and waited there for about ten seconds until it disappeared. This was Keri's first visit to the camp and there was no Bigfoot activity that we were aware of — not unusual for any newcomer. After I went to sleep, Keri and another gentleman who we had with us heard some tree knocks or 'pops', from about thirty yards away.

One other time in camp, during the 70s, when I was with Bill, my horse-packin' buddy, we witnessed a controlled ball of blue light coming down from the sky. It was like the moon was falling in slow-motion. We lost track of it behind the tree line about one mile away. We didn't hear any sound from it, but it was pretty interesting to see. We talked about it—not much else to do but call it a UFO.

Is There a UFO Connection with Bigfoot?

As previously mentioned, Al Berry, the investigative reporter who came looking for a hoax asked me to not relay our unusual stories — strange stories like this would put our credibility in question. That was several decades ago, but now I just don't really care — I saw what I saw.

Al and I eventually became good friends and we traveled together a lot. We were adopted into the BFRO (bfro.net) as Curators, investigated reports, met with interesting people, and spoke at conferences. He also taught me how to fly-fish, something he was really good at it.

Kern Peak, California: Al came across a book that went out of print in 1934. In this book was a picture of a huge five-toe print alongside a human print. What made it interesting was that both prints were embedded in granite on top of Kern Peak, just below Mt. Whitney in the Sierras of California. Something that made that area very attractive to me was the dirt airstrip below that mountain in a meadow 8,700 feet above sea level — did I mention I had an airplane? Another interesting tidbit was that for years UFO's have been spotted in that area.

We flew there, pitched camp, and the next day began to hike up to the top of that mountain looking for the granite rock with those embedded prints. We spent three days there hiking, looking intensely, and fly fishing. Except for the fish that Al caught, we came away empty-handed — no prints were found and no UFO sighted, but we sure had a lot of fun.

Once Al and I flew to northern California to interview a guy who claimed multiple Bigfoot sightings on his property. However, at this point in time, it was a red woo-woo flag for us when entering someone's home to see lit candles all around the place. How those places kept from burning down I'll never know, but the place seemed nice and warm.

Is There a UFO Connection with Bigfoot?

I understand those people better now, but I didn't then — it seemed strange. When some of those people told us about seeing a UFO, I didn't act indifferent, but after so many of those reports, I had to give a little jaw-drop and raise an eyebrow. Something was going on — something out of the norm — too many people reporting this kind of stuff.

Megalith Gate at Ollantaytambo, Peru
Author: 2014

Is There a UFO Connection with Bigfoot?

After so many years of hearing reports like that, traveling the world, and seeing the unusual artifacts in South America, I've come out of the closet — so to speak, and have drawn a conclusion. First, without a shadow of a doubt, aliens have visited this earth and, second, from the skeletal remains in Peru apparently copulated with humans.

The team's archaeologists (I was with two different teams on two different expeditions.) did analysis of these skulls. As pictured on the previous page, the pre-Inca culture that was in Peru and Bolivia built structures that continue to baffle scientists. Boulders weighing well over 100 tons were moved from miles away and placed on a thirteen thousand foot mountain top. How could any culture do that? Thousands of men pulling those boulders with ropes just didn't seem logical.

It has been suggested they were able to tune into the vibrational frequency of those boulders and move them with sound frequency, but the bigger mystery is how they cut them so precisely and placed them together without mortar — like a puzzle. The Incas didn't have tools that could do this. It's been proposed that the reasoning for the puzzle-like cuts was for earthquake stability, but the 'how' is still another question.

As mentioned, one visit wasn't enough — I went to Peru twice and on into Bolivia. I was privileged to be in the company of Brian Foerster, a man who is considered one of the most knowledgeable about different cultures from ancient times and really knows his way around that entire area.

Once we were escorted by the Chief of Police, along with armed guards, to a 7-mile long cemetery which had a deteriorating pyramid. We stood atop the pyramid and viewed a group of trees in the distance (the only trees around). The chief said a few years back there used to be a lake there, but a UFO came, hovered over it, and when it left the lake was dried up.

Sillustani, Peru: Note below: cut boulders, perfectly rounded with no mortar. Built by a pre-Inca advanced culture. What these were used for is a mystery.

Author: 2014, Peruvian Expedition

The Star Child

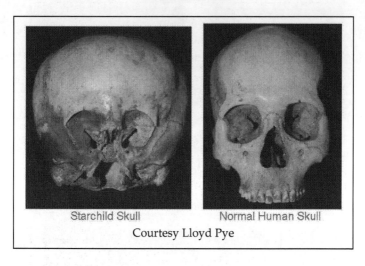

Courtesy Lloyd Pye

UFO's are commonplace in South America, but also in Central America. The Star Child found in Copper Canyon, Mexico, is debatably of alien origin. At a conference in Bellingham, Washington, in 2005, I had the privilege of being one of the speakers along with the late Lloyd Pye. (lloydpye.com)

You may recall Lloyd Pye was the man who spearheaded the study of the Star Child skull (starchildproject.com) almost two decades ago. Mr. Pye gave some very convincing reasons why the remains were not a product of congenital hydrocephalus and after listening to Mr. Pye's presentation, I was left with little doubt that it was indeed an alien skull.

Is There a UFO Connection with Bigfoot?

An Example of Natural Cranial Elongation

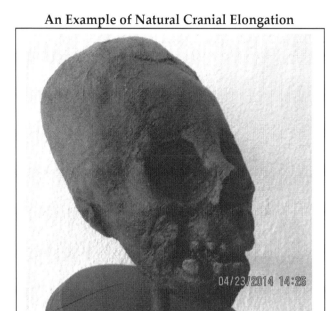

The Peruvian Paracas infant skull.

Deduction

So with the many reports of Bigfoots being seen in conjunction with UFO's, some reportedly being beamed down from a craft, it should give us all a little pause. If Bigfoots are the product of alien intervention or being influenced by aliens, what would that mean? Why are they here, hiding, watching, and waiting?

The list of people who have claimed an encounter, or even abduction, by aliens is too long to mention. I've found very few people who think that UFO's are fictional. To those few people, I say, "Think about it, think of the untold numbers of galaxies and stars in the universe."

Is There a UFO Connection with Bigfoot?

I think scientists acknowledge there may be other species, but if there are no specimens, 'they' don't exist. I think it's completely naive for anyone to think that in this vast universe we are the only life-form entities that exist.

If these beings called Bigfoot are a remnant of celestial intervention into earthly beings, we, as humans, need to be mindful and use our discernment if we are approached by one of them. For, what or who may have given them sapience we don't know.

Artist: "Sev7en,".

Chapter 17
Researchers and Researching

"I think, as serious researchers, it is important to be mindful, to not discount people who share experiences outside of our body of knowledge. Yes, we must be discerning, but open disbelief implies they are being dishonest. And for 'All' we know — we just don't know."

Keri Campbell, quote from Kelso, Washington 2017 Bigfoot Conference

Keri Campbell afield

Researchers and Researching

To research these beings known as Bigfoot we should use all the tools in our scientific arsenal. We look for any and all the logical answers using classical science. We don't try to make something out of nothing. If a researcher hears a certain sound that he or she can't identify, that doesn't mean that it came from a Bigfoot.

> "Hearing an unusual yell or scream while in the woods is common, it does not mean it was a Bigfoot — unless you actually saw a Bigfoot make that sound." Peter Byrne

Two hundred years ago if you told someone that we could communicate with others from across the world instantly, or even tell them they will be able to fly across this country in a few hours, they would have considered you crazy. Less than five hundred years ago you would have probably been burned at the stake. Technology is changing rapidly and the flow of information is accelerating at an unusually fast pace. What is considered common today will be antiquated tomorrow. So again, I suggest we all open our minds to possibilities that may not fit our current beliefs.

For some people nowadays it's still a question of whether or not Bigfoot actually exists. After 1971 that was not a question for me — I had experienced them firsthand, and that took my awareness to a different level. There was some kind of big unknown animal in those woods. When I first got involved with the research of these beings my position was that they had to be an undiscovered giant ape, and for years it remained my only position. At that time I had never delved into quantum physics. Being a Christian, however, I looked into the history of giants and how they came into being from a biblical standpoint but also figured those days were long gone.

Although we had unexplained happenings when dealing with these beings at our *Sierra Camp*, I figured there were probably other reasonable answers that would somehow explain those events. For years afterward, however, I listened to other people talk about their encounters and I thought some of those folks have a very creative imagination. Then it suddenly occurred to me, "Could these giants represent something else? Or are they really just a giant ape?" Perhaps like so many Native Americans claim, "a Spiritual Being of some kind?" Here one minute, and gone the next.

In that same camp, we have the Bible carrying folks that claim these giants are the Nephilim, from the Good Book of Genesis, and who will come again in the last days. So over the years, I've concluded that there are two camps in this world of Bigfootery. We have the Apers; researchers that think of these beings as no more than an undiscovered ape that bleed, eat, and poop, just like any other animal. They can be tricked, trapped, shot or somehow caught; it just takes time.

But then we also have the Woo-Wooer's who think these beings can change dimensions by passing through portals, read minds, or set informational sticks on the ground, so they'll know how to get where they want others of their brethren to go. Who's right?

I've met a lot of sincere people in my years of research and I'm sure that they think their experience makes their belief right. Could the correct answer lie somewhere in-between the Apers and the Woo Wooer's? Could they be flesh and blood and still have abilities that we just need to understand?

In my opinion, we humans have abilities far above what we now display. Scientists say that we don't use even 10 percent of our brain. Does that mean we can evolve into something better? Could these beings we know as Bigfoot, or Sasquatch, have evolved into something higher than just an ape? Or somehow been given an enhanced attribute that we have yet to understand?

Just as we physically grow from a childlike state, our idealistic hypotheses can, and should, change with information and experience.

> "An effective researcher does not selectively search for reports that fit his or her paradigm. None of us actually know, so a good researcher must keep an open mind. We search and research for Truth and that Truth will defend itself."

Ron Morehead from his Sasquatch Summit lecture 2016

Dos and Don'ts

I've certainly changed many of my thoughts over the years and want to share what has worked for me. I've learned to broaden my scope and heighten my awareness. Here are suggested Dos and Don'ts that have worked for me:

Do
- Be very clear with intentions
- Locate remote 'Hot Spots' near water, i.e., streams
- Small Groups (2 - 4)
- Habituate the area
- Be predictable
- Relaxed atmosphere
- Be Sincere and sober
- Create curiosity — have fun
- Offer them food, good food
- Horses and guns are OK
- Know forest animal sounds (see endnote)
- Listen for the sound of silence

Don't
- Don't be the aggressor
- Don't be fearful
- Don't take dogs
- Don't shine flashlights
- Don't jump around when you hear unidentified sounds
- Don't try to trick them

I suggest documenting what you do and share that knowledge and experience with others. Use your discernment and don't underestimate them. Be prepared for anything...you won't find them — they will find you. Somehow Bigfoot is able to entangle humans, quantumly.

Endnote: I recommend on-line study of the Cornell University Department of Ornithology Macaulay Library of animal sounds. The Library has cataloged over 150,000 verified animal sounds. For example, a study of owl sounds may surprise most field researchers.

Chapter 18
About Einstein

"It is intuition that improves the world, not just following a trodden path of thought." Albert Einstein

Albert Einstein has taken, or been given, credit for several discoveries relating to physics that have been questioned. It's said that other physicists discovered much of the quantum problems that he wrote equations for without credit given to those scientists. That being said, Einstein still has the most recognizable equation of all time $E = mc^2$. However, the reader should know Einstein's famous equation has firm roots in the history of physics as the following quote from Moody indicates.

> "The conversion of matter into energy and energy into matter was known to Sir Isaac Newton ("Gross bodies and light are convertible into one another...", 1704). The equation can be attributed to S. Tolver Preston (1875), to Jules Henri Poincaré (1900; according to Brown, 1967) and to Olinto De Pretto (1904) before Einstein. Since Einstein never correctly derived $E = mc^2$ (Ives, 1952), there appears nothing to connect the equation with anything original by Einstein. This actually substantiates the correctness of the equation."
>
> Richard Moody, Jr., excerpt from Nexus Magazine

"Though he was the most famous scientist of his time, Albert Einstein knew we could never fully understand the workings of the world within the limitations of the human mind. Experiencing the universe as a harmonious whole, he encouraged the use of intuition to solve problems, marveled at the mystery of God in nature, and applauded the ideals of great spiritual teachers such as Buddha and Jesus."

Richard Moody, Jr.in stage.upliftconnect.com

In Search of the Cosmic Man

The following is an excerpt of his writings that explores the meeting place between science and spirituality, giving us a fascinating glimpse into how Einstein saw the world: Text Source: *Einstein and the Poet: In Search of the Cosmic Man* (1983). From a series of meetings William Hermanns (d: 1990) had with Einstein in 1930, 1943, 1948, and 1954.

"School failed me, and I failed the school. It bored me. The teachers behaved like Feldwebel (sergeants). I wanted to learn what I wanted to know, but they wanted me to learn for the exam. What I hated most was the competitive system there, and especially sports. Because of this, I wasn't worth anything, and several times they suggested I leave.

This was a Catholic School in Munich. I felt that my thirst for knowledge was being strangled by my teachers; grades were their only measurement. How can a teacher understand youth with such a system?

Order in the Universe, Disorder in the Human Mind

"From the age of twelve, I began to suspect authority and distrust teachers. I learned mostly at home, first from my uncle and then from a student who came to eat with us once a week. He would give me books on physics and astronomy.

Einstein as a youth

The more I read, the more puzzled I was by the order of the universe and the disorder of the human mind, by the scientists who didn't agree on the how, the when, or the why of creation.

Then one day this student brought me Kant's *Critique of Pure Reason*. Reading Kant, I began to suspect everything I was taught. I no longer believed in the known God of the Bible, but rather in the mysterious God expressed in nature.

The basic laws of the universe are simple, but because our senses are limited, we can't grasp them. There is a pattern in creation.

If we look at this tree outside whose roots search beneath the pavement for water, or a flower which sends its sweet smell to the pollinating bees, or even our own selves and the inner forces that drive us to act, we can see that we all dance to a mysterious tune and the piper who plays this melody from an inscrutable distance—whatever name we give him—Creative Force, or God—escapes all book knowledge."

Unveiling the Magnificence of Creation

"I like to experience the universe as one harmonious whole. Every cell has life. Matter, too, has life; it is energy solidified. Our bodies are like prisons, and I look forward to being free, but I don't speculate on what will happen to me.

I live here now, and my responsibility is in this world now. I deal with natural laws. This is my work here on earth.

The world needs new moral impulses which, I'm afraid, won't come from the churches, heavily compromised as they have been throughout the centuries.

Perhaps those impulses must come from scientists in the tradition of Galileo, Kepler and Newton. In spite of failures and persecutions, these men devoted their lives to proving that the universe is a single entity, in which, I believe, a humanized God has no place.

The genuine scientist is not moved by praise or blame, nor does he preach. He unveils the universe and people come eagerly, without being pushed, to behold a new revelation: the order, the harmony, the magnificence of creation!

And as man becomes conscious of the stupendous laws that govern the universe in perfect harmony, he begins to realize how small he is. He sees the pettiness of human existence, with its ambitions and intrigues, its 'I am better than thou' creed.

This is the beginning of cosmic religion within him; fellowship and human service become his moral code. Without such moral foundations, we are hopelessly doomed."

Improving the World with Ideals, not Scientific Knowledge

"If we want to improve the world we cannot do it with scientific knowledge but with ideals. Confucius, Buddha, Jesus and Gandhi have done more for humanity than science has done.

We must begin with the heart of man—with his conscience— and the values of conscience can only be manifested by selfless service to mankind.

Religion and science go together. As I've said before, science without religion is lame and religion without science is blind. They are interdependent and have a common goal—the search for truth.

Hence it is absurd for religion to proscribe Galileo or Darwin or other scientists. And it is equally absurd when scientists say that there is no God. The real scientist has faith, which does not mean that he must subscribe to a creed.

Without religion, there is no charity. The soul given to each of us is moved by the same living spirit that moves the universe.

I am not a mystic. Trying to find out the laws of nature has nothing to do with mysticism, though in the face of creation I feel very humble. It is as if a spirit is manifest infinitely superior to man's spirit. Through my pursuit in science, I have known cosmic religious feelings. But I don't care to be called a mystic.

I believe that we don't need to worry about what happens after this life, as long as we do our duty here—to love and to serve.

I have faith in the universe, for it is rational. Law underlies each happening. And I have faith in my purpose here on earth. I have faith in my intuition, the language of my conscience, but I have no faith in speculation about Heaven and Hell. I'm concerned with this time—here and now."

It is Intuition which Advances Humanity

Many people think that the progress of the human race is based on experiences of an empirical, critical nature, but I say that true knowledge is to be had only through a philosophy of deduction. For it is intuition that improves the world, not just following a trodden path of thought.

Intuition makes us look at unrelated facts and then think about them until they can all be brought under one law. To look for related facts means holding onto what one has instead of searching for new facts.

Intuition is the father of new knowledge, while empiricism is nothing but an accumulation of old knowledge. Intuition, not intellect, is the 'open sesame' of yourself.

Indeed, it is not intellect, but intuition which advances humanity. Intuition tells man his purpose in this life.

I do not need any promise of eternity to be happy. My eternity is now. I have only one interest: to fulfill my purpose here where I am.

This purpose is not given me by my parents or my surroundings. It is induced by some unknown factors. These factors make me a part of eternity."   Albert Einstein

Chapter 19
Al Berry

Al Berry 1941-2012

Al Berry came to the *Sierra Camp* in 1972 feeling sure he would uncover a hoax. I had started going there with my friends the previous year. Unbeknownst to all of us, during our hikes, he was rummaging through our supplies, backpacks, and storage barrels looking for signs of anything that would lead him to the culprit.

I wanted to like Al and tried hard not to make a snap judgement, but at first, it was difficult. He was always questioning us, more like interrogating us, about what we were doing. At one time or another he actually accused each of us of perpetrating these goings-on, but obviously without the outcome he'd hoped for. At that time I didn't understand why he found it so difficult to accept what we believed to be real.

It wasn't until a few years afterward that he and I became close friends and began researching, interviewing, and speaking at events together. It was 1976 when he co-authored the book, *Bigfoot*, with Ann Slate. He bummed me out a bit by misspelling my name. The first three chapters in that book are about our Sierra Camp and a handful of the strange happenings that went on there. By then he'd accepted that the events were probably real and that the camp still held an unusual mystery. On occasion he would tell me that he thought it would have made a better story if it had been a hoax. How anyone could have pulled this off without leaving a trace was something that bothered him for years.

Al had a Masters in science, and also had a degree in the Kings English. He was an excellent writer and when I fell short in that area he ended up coaching me, often to my dismay, but actually, it was good for me. Also I noticed if anyone got on the wrong side of Al's pen that person was in deep horse hockey.

Over the years he and I became very close, ventured into a lot of beautiful country, spent a lot of time around a campfire, and exchanged a lot of ideas about the creatures we'd experienced. Although there were several mysteries still being held captive, having a background in science Al was having a hard time buying into anything paranormal. And as I stated earlier, he advised me to not speak of paranormal issues in public because I would lose credibility.

Years later, with Al's warning in mind, I began to delve into the multi-dimensional aspects of quantum physics. I can't even recall the exact article or conversation that sparked the match over a decade ago, but soon a full bonfire was burning in my brain!

Anyone can claim a story like the Sierra Camp, but what gives it credibility is the science behind the data, not just someone saying that it happened. What all of us hunters obtained were recorded vocalizations of these creatures mouthing off; with no idea of what was being said or why.

Al arrived at our Sierra Camp in 1972 with recorder in hand. It took a few trips but once he got something of his own on tape, he immediately began to seek out professional help; help that would catapult this amazing evidence to the forefront of the public's eye, or so he thought.

In early November 1973, after months of attempting to locate and enlist appropriate scientific aid in the analysis of the Bigfoot tapes, Al phoned I.E. Teibel, president of Syntonic Research, Inc., in New York City. Teibel recently had appeared on a nationally syndicated television talk show discussing former President Nixon, the Watergate tapes, the relative ease by which tapes can be altered, and the relative difficulty in detecting such alterations.

On the talk show Teibel stated his firm was expert in exposing hoax tapes. Therefore, Al thought it seemed reasonable to expect that through him he might get to the bottom of the creature sound mystery; for a fee, anyway.

Teibel agreed to listen to a copy of the sounds, and on November 11, 1973 Al sent a copy to him. In a cover letter, Al wrote:

"I understand that you can't make any guarantee about proving or disproving {the tape's} authenticity. My hope is that you can help me build a case one way or another, however, or at least provide some basis for probable best interpretation where there remains a question mark..."

Less than a week later, on November 16th Teibel wrote back:

> "We have examined the tape copy you sent us on November 11th, utilizing the highest quality ¼ -track stereo playback equipment obtainable. A spectrum analyzer and a dual trace oscilloscope were also attached in the signal chain so that visual analysis could be obtained simultaneously. Our findings are as follows…"

The sounds were undistorted, he said. He also said and suggested that because of the power necessary to transmit low-frequency sounds, the source could not have been more than thirty feet away from the receiving microphone. He also said, the volume and "color" of the sounds remained relatively stable which indicated the creature, assuming it was a creature, had not moved about while producing the sounds.

He further said that the sounds were "almost completely localized on the right channel" (the original recording was in stereo), which could only happen if the tape recorder's transducer malfunctioned or "the sound source was very near the microphone" (Teibel's emphasis). He added that there were "unmistakable" mike-handling noises and indications of tape-drive interruptions that led him to believe the recordings were not continuous in nature.

"Since we do not have access to the original recordings," Teibel continued, "our findings are based on the copy provided…"

Al had not forwarded the original tapes pending Teibel's report on the tape copy. Until their probable value was ascertained, it seemed foolish to risk their possible loss or destruction, either in the mail or as a result of careless handling. After reading Teibel's report, Al knew that Teibel was in error about the continuous nature of the recording. And there was more to come in regards to the channel on the right picking up most of the sound. It was years later, 1996 to be exact, while producing Volume I of the Bigfoot Recordings in a Burbank, California studio that we found out it was probable Al inadvertently secured his microphone in a vertical position on the small tree with the microphone's left side facing toward the ground.

On November 20, 1973, Al mailed two of his original cassettes to Teibel, along with a letter detailing the time, place, and circumstances under which the recordings were made:

> "From your analysis," he wrote, "it seems possible that one of the tapes (on the two dates in question, Oct 21 and 22, 1972, separate recordings were also made by Warren Johnson, Bill McDowell, and Ron Morehead that were partial and noncontinuous in nature, and may have been a master pre-recorded elsewhere then amplified from somewhere in the immediate vicinity out into the woods. I want to explore this possibility ... It was the only possibility that seemed plausible to me, given Teibel's preliminary findings: and if the evidence were truly there, on the original tapes, it would be a clear-cut indictment of Warren's group, and the authors' gullibility."

Al then offered the Johnson/McDowell/Morehead tapes to Teibel as soon as they could be obtained.

In yet another letter, Al provided more details about the camp scene and circumstances. Perhaps out of frustration, and realizing more than a year's investment in time, effort and research of the phenomenon was at stake, Al added a paragraph about the night Larry Johnson and he had listened to the Old Man (Bigfoot) sing his swan song.

> "The point in drawing this picture for you is, would a hoaxer subject himself to such conditions all night long — in a remote wilderness setting, without knowing or having any way to know whether the persons he was hoaxing were awake and paying attention?
>
> Try to imagine yourself there as I was; half frozen in a crude lean-to shelter, in fairly dense woods, at about 8,500 feet in late October after the first snowfall, some 2,000 feet higher than the nearest road and about eight miles distant from the nearest established trail. Is this setting alone constant with the idea of a hoax, either inside or outside this small group of people which had been documenting similar experiences nearly to the point of boredom for more than a year before I met them?"

On December 6, Al received a letter from Mike Kron, one of Teibel's research associates, with remarks pertaining to the original recordings:

> "We have played both tapes ... in my opinion the sounds recorded have been spontaneous in nature and seem to have taken place at the time of the original recordings. There are some questionable clipped delay times at the end of certain utterances, and the recordings themselves are not of the highest quality, due to the extreme conditions under which they were made and the technical limitations of the microphone and tape recorder utilized. This thereby limits the amount of analysis that

can be performed, as we do not have a standard reference model for comparison, as would be the case with voice-print spectrograms.

I must state that the frequency range of the utterances could be considered humanoid and that these sounds could have, theoretically at least, been produced by a human adult male. We are not saying that this is the case, however.

Tape recordings are extraordinarily flexible and subject to all sorts of artful manipulations. The fact that a photograph was not made of the beast supposedly on this cassette recording limits the credibility of the recordings.

If you wish us to prove 100 percent that such a recording is either "fake" or "genuine" we cannot do either. Just as a recording of a "Martian" could only be analyzed by speculation. We have attempted to contact a zoologist here in New York who might wish to comment on these recordings but has been unable to do so as yet....

I would suggest approaching a nearby engineering university for the detailed analysis these tapes require."

Teibel's Assessment

Kron's letter was accompanied by one of Al's own, sent earlier to Teibel, in which Al presented a number of specific questions. Here are Teibel's remarks in response to Al's questions.

Al: "Had the Bigfoot sounds come from a hidden amplifier?"

Teibel: "Interchanges are interesting—seem spontaneous—no obvious defects in sound although it could be from a speaker."

Al: "Was there any sixty-cycle "hum" evident, which would indicate indoor prerecordings?"

Teibel: "No hum evident from an external source."

Al: "Was the source in fact "very near" the microphone, as suggested by the preliminary analysis of the duplicate tape?"

Teibel: "Equalization in re-recording (the copy Al had initially sent) changed emphasis somewhat. Original (tape) does not seem to indicate close recording."

Al: "What was the frequency range of the sounds?"

Teibel: "Basic freq. 200-2K Hz. Screeches much higher (5K)."

Al: "What is a human's range?"

Teibel: "Approximately same."

Al: "What is a gorilla's range?"

Teibel: "Lower range—1 more octave."

Teibel's final comment: "Get a picture!"

On December 14, 1973, Al received a final letter from Teibel:

"We have attempted to interest various local scientists in examining the tapes you have provided. To this date, we have been in correspondence with several people who we thought might have an interest in pursuing this matter on the basis of scientific interest alone. As we have been unable to obtain any positive response, and a week has passed, we are herewith returning the four tape cassettes (including one from Warren and one from Ron that Al had sent) you have provided…

My suggestion as to your further course of action would be to contact scientists closer to your locality, as verification based on tape analysis is extremely fragile, especially in light of the limitations of most tape cassette recording equipment and the absence of corroboration, either by photographic means or an expert witness.

Once you meet such a person, we feel you will be able to provide spontaneous data which will be helpful in examining this material and other material you might have. I would also suggest the following chain of activities: have a zoologist review the tapes in your presence; have him comment, in writing, on his observations; then submit the recordings and as much corroborative material as possible to a university acoustics laboratory which might undertake a thorough examination on the basis of scientific interest." I.E. Teibel

Al Berry

Al Berry (in lead) and Ron Morehead riding to Sierra Camp. Photo by Al's son Greg Berry.

Al was 6 foot 4 inches tall and his intelligence was taller. He was an officer in Vietnam and saw plenty of action. My experience of Al was that of a perfect gentleman.

Al was always trying to get to the bottom of the occurrences at the Sierra Camp — he told me, more than once, that he "wished it was a hoax — easier to understand," he said. We spent many of our days discussing the 'What ifs', and he always kept an open mind, but because of his scientific background had a difficult time digesting the unexplained happenings. Honestly, if it weren't for Al's curiosity and tenacity, all of the in-depth vetting of the *Sierra Sounds* would not have happened and for that I owe him a debt of gratitude.

Chapter 20
On Dolphins and Bigfoot

Joan Ocean

Joan Ocean is currently regarded as a leading authority on the subject of Dolphin Tel-Empathic Communication. She developed the methodologies of her work, entitled "Participatory Research," in which human and cetacean species are equally conducting research with each other, and has dedicated her life to studying cetaceans by respectfully joining them in their natural habitats and becoming their friend.

Author's note: Joan Ocean wrote the following article expressly for *The Quantum Bigfoot*. I sincerely thank her for that. Joan's real life experience is everything this book is about.

On Dolphins and Bigfoot

There is More Going On than Most Know

Joan Ocean, MS
Hawaii 2017

"When I first met the free-swimming dolphins in the warm waters along the Kona Coast of the Big Island, I was intrigued by their behaviors. They obviously responded to each other as family members would, swimming side by side, diving in unison, communicating with many different whistles and squawks, mating often and giving birth to their young who they obviously loved and cared for. They seemed unconcerned about my close presence. They carried on in their daily life as if I wasn't there.

And so I continued to swim alongside them for months that turned into years, getting to know them by their identifying marks on their bodies, tails, and fins. Feeling honored by their willingness to have me alongside them, feeling like one of their family members ...of course, I could not dive as deeply as they can and I couldn't stay with them if they decided to swim fast and move on to a different bay for the day. Then I would say goodbye and hoped they would return again the next day.

And they did! Day after day, they swam into the bay where I lived in a little house along the shore, and upon their arrival, they would leap into the air to be sure I could see them. I would immediately go into the bay and swim toward them. They would come toward me and we would close the gap between us.... turning into each other, to swim side by side. There was no mistaking their behavior. It was such a wonderful and satisfying feeling to see that they remembered me and seemed as happy to see me as I was to see them.

As a youngster, I had had Rheumatic Fever and was restricted from involvement in any sports or athletic activities. And so this new behavior of mine, swimming with them was exhilarating for me. I felt strong in the ocean and safe too when they were by my side. I noticed that they came to me allowing me to swim in their slipstream or water-flow-field as a way to keep me alongside the rest of the pod. If I fell behind while they circled wide, round and round in the bay, one or two would double back to swim alongside me and help me catch up

At some point, I had the ah-ha moment where I realized they were accepting me as a member of their group. And my heart overflowed with love, as I acknowledged their kind hospitality and relaxed into their gentle company.

More and more revelations came into my mind as I escorted them around the bay each day. They were sending me information about themselves and about their ocean home. Back on land I dutifully wrote it all down and kept it in a journal. My enthusiasm to be with them accelerated to the point where I did not want to go anywhere else, no vacations, no trips to the mainland, no shopping in town -- I only wanted to be with them and learn from them. I called our time together: Participatory Research. This was
something we were doing together. Teaching and learning....... I began to feel they wanted me to know them and to tell others about them -- who they really are.

I facilitated a Conference in Kona, Hawaii about the dolphins and because I often like to invite a speaker who is different from the rest, I chose to invite a man who had written a book about Bigfoot. It seemed that some of his experiences of communication with them were similar to my growing communications with the dolphins. This led to my going into the forest on the mainland for a week, to see if there really were any humanoids by the name of Bigfoot or

Sasquatch. I was doubtful but curious since many respected and sincere people were claiming proof of their existence.

Contact in the Kiamichi

During that first week, alone in the beautiful, remote Kiamichi Mountains, I was enjoying warm Fall weather with the door of my log cabin open to view the autumn leaves and the setting sun. Sitting on the cabin floor, I looked up when I felt a shadow pass over me. I was shocked and then gratified to see a tall hairy Sasquatch silently standing outside my cabin door. Unmoving, she stood there, then lowered her head, leaning down to look inside the doorway at me.

Returning her gaze, making eye contact, I felt a sense of peacefulness come over me. I had no fear. Her eyes were kind and loving and gazing directly into my eyes. By all that is rational, I should have been scared or at least surprised, but no — I felt calm and even a timeless feeling of contentedness in her presence.

Did we connect for a minute or was it for an hour? I was soon to learn that our relationship to "time" can be different in the world of these Sasquatch. In addition, I have learned that they can emit by choice, a frequency of calmness or a frequency of fear into their surroundings as a protective measure when needed.

We remained gazing into each other's eyes, motionless until this seven-foot female Bigfoot lifted her head and silently lumbered away. I did not move. I was truly spellbound. I felt surrounded in feelings of gentleness and harmony; I didn't want to lose it. My mind was racing to understand what I had seen, but my feelings were of a deep contentment, a knowing, even a familiarity with this hairy humanoid.

My feelings confounded me more than the experience itself. Shouldn't I have been afraid? Alone here with no one else around?

Shouldn't I have backed away from this massive hairy thing? Shouldn't I have yelled out in surprise?

No....I felt honored, I felt relaxed and peaceful, I felt extremely grateful and quietly elated. Inexplicably, I felt love. I had come to this place to meet them and this one had come to meet me.

And so this began my friendship with a family of 33 Sasquatch who lives in the forest.

As the years passed and I spent weeks in the Spring and weeks in the Fall with them, I learned they had many of the same abilities and advanced sensory perceptions that the dolphins also exhibited.

Because of the experiences I have had with the dolphins (*Dolphins into the Future*, published 1997), I found I could communicate very naturally with some of the Sasquatch family as well. They seem able to project the sound of their voice so that it is not always evident where they are coming from ... in front of me, behind me, up on the mountain, across the river? I learned to enjoy their multiple sounds, some of which closely mimicked human words, and I didn't need to look this way and that to see them.

They are excellent at telepathic communications whenever I open myself to commune with them in that way. Often I am having an unspoken thought when I sense the warm feeling of their presence and a response comes into my mind in answer to my thought. Many of their communications surprise me with their clever understanding of the human race. And it is more than a thought. For example, when I asked them one day where they were since I hadn't heard from them the night before, they answered "We at South Lake."

When I receive a transmission from them, it is not just words, in fact, there are very few words. I receive an image. In this communication, I was suddenly immersed in a vision of them in a lake, cooling off, and in joyful company with some of their family members. Their relationship to Nature is very profound and is part of their daily life. When I receive information like this, it seems to also carry their feelings, their physical contentment (enjoying the cool water on this very warm day) and the beauty of the surroundings where they are. I receive impressions in my mind of their location, without having been there. It is as close as you can get, to being there yourself.

This is also the method of communication I experience with the Hawaiian Spinner dolphins. While swimming in the ocean with them, they send me sound pictures that I have named: Acoustic Images. They communicate with each other this way and also with me. Rather than telepathy, I call it Tel-Empathy. It is more than a mind-to-mind transmission of thoughts, it includes feelings and images. Communicating non-verbally is a much more efficient and complete method than using only words and mental processes.

One evening I asked their permission to bring a few people to meet them if I develop a protocol for their safety and anonymity. Eventually, they said, yes. I interviewed respondents thoroughly, but occasionally someone came along, who had fear and doubt. A little of that is a good thing, but not if it would make the person overly excitable or nervous. Then the Sasquatch would hold back in coming close. Whereas normally I would hear them approaching, walking in the crunching dead leaves with easy long strides and then they would stand behind us for a distance or sometimes very close depending on how accepting the people were.

Always wanting to see them, I would ask if they would come forward. But they responded that experience had taught them that the human race was frightened by their appearance and would become unpredictable (perhaps hurting themselves) due to fear. And that they preferred to make contact in the dark and with people who were ready. They could tell somehow if we were ready. And their contact with us proceeded accordingly. Most everyone who joined me had an unusually strong desire to meet the forest people and befriend them. They were not disappointed.

Whenever I arrive in the forest, I inquire where they are. And their answer is always the same: "We are here." I look around and I don't see them: " Where? Where are you?" As time went on, I began to realize that they were right. They are always here. This was of great fascination for me as I began to understand their ability to transcend time and space. They seem able to drop-in to whatever space, at whatever time, they want to, to be with us. Somehow our limiting linear-time beliefs are not applicable to them.

The dolphins had already communicated to me; those dolphins are much more than their physical bodies and because of that, they can enter dimensions that humans are unaware of. Because of this revelation from the very wise and inspiring dolphins, I could also understand the similar quantum field of consciousness of my new Sasquatch friends.

Through acoustic imagery, the dolphins revealed to me a process called 'vertical intervention' or 'a window in time'. I soon learned that these particular Sasquatch people were also capable of experiencing this phenomenon of coming and going, slipping in and out of time.

When I questioned them about Portals, they referred to these windows in time, as the "Sometime Place". When I then questioned their obvious ability to appear and disappear as needed and referred to it as 'de-materialization', they explained it differently, saying it was a temporal timeline they can enter.

The scientists exploring wormholes and the new advanced physics of today might call it: traversing parallel timelines. Apparently, we are experiencing the merging of temporal timelines now. These friendly Sasquatches are already demonstrating this ability to alter reality via their active, expansive consciousness.

Molecular physicists know that in Quantum Physics electrons can pass through solid matter. Similarly, while in a quantum state, the Sasquatch have no weight or mass; they are like a "wave" of energy, or perhaps like an Orb. Can a human body, with all of its electrons, move through physical barriers as well?

This is what modern scientists are investigating in advanced laboratories around the world. I believe the dolphins, and other life forms on Earth such as the Sasquatch, can do this - so why not humans? Perhaps this is what we shall learn from our gentle Sasquatch friends once we earn their trust, begin to commune with them and seek to protect them and their homes, the natural environments of planet Earth."

End of Joan Ocean article.

Chapter 21
Albert Ostman's Account
Kidnapped by a Bigfoot

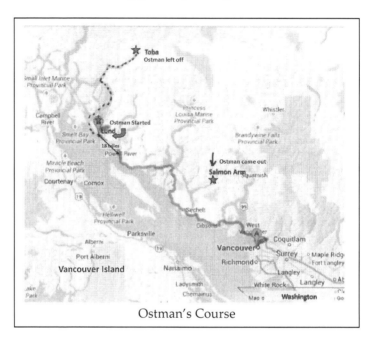

Ostman's Course

This chapter is about Albert Ostman (circa 1893-1975) who claimed he was kidnapped and held captive by a Bigfoot. On the surface, the story seems to have little to do with the quantum aspects of these beings. But after compiling and reviewing the previous 20 chapters of this book, the kidnapped man's account may very well relate to quantum physics. So, I think my research into his story should be told.

First, let's consider the source of the original account and my summary of it. John Green (1929-2016) was one of the most respected Sasquatch researchers in the world. His books were among the

Ostman's Account

earliest, and are foundation stones in the literature. In Green's monumental work, *Sasquatch, The Apes Among Us,* (Cheam Publishing, 1978) he reports a 1957 interview with Albert Ostman, who said, in 1924 he was kidnaped and held captive by a Bigfoot family.

From *The Apes Among Us*:

> "Albert Ostman is dead now, but I enjoyed his friendship for more than a dozen years and he gave me no reason to consider him a liar. I have had him cross-examined by a magistrate, a zoologist, a physical anthropologist, and a veterinarian, the latter two being specialists in primates....all sorts of skeptical newsman (sic) have grilled him....none was able to trap him or discredit his story....although the magistrate in particular tried very hard to give him a rough time."

From Wikipedia: On August 20, 1957 Canadian police magistrate A.M. Naismith wrote an affidavit on Ostman, and his story, which states;

> "...I found Mr. Ostman to be a man of sixty-four years of age; in full possession of his mental faculties. Of pleasant manner and with a good sense of humor. I questioned Mr. Ostman thoroughly in reference to the story given by Mr. Green. I cross-examined him and used every means to endeavor to find a flaw in either his personality or his story, but could find neither..."

Albert Ostman also signed a Solemn Declaration indicating that his account of the Sasquatch story was true under oath and by virtue of the Canadian Evidence Act.

Ostman's Account

The interview John Green conducted has for years been considered an iconic story, and in the minds of many people verifiable because of the many details Ostman included; details he could only gain from first hand observation. However, as valid as the account may be, some of the facts do not add up.

After 33 years from the time of the incident in John Green's interview, John wrote the following — as given him by Ostman — and summarized by me:

- Ostman would not swear to the accuracy of his trip, other than what happened during and after his abduction.
- He was on vacation from construction, looking for a lost gold mine.
- He took a Union Steamship to Lund, B.C. From Lund he hired an elderly Indian to row him to the head of the Toba Inlet, a distance of not less than 50 miles.
- After arriving, the same day Ostman hiked into the mountains where, from about 1,000 feet above sea level, he saw boats in an inlet.
- Ostman hiked northeast from the inlet for about 10 miles into the wilderness. He shot a deer and set up camp. At night something began milling around, but he figured it was a porcupine. So, he placed some of his belongings inside his sleeping bag.
- That night he was picked up in his sleeping bag and carried for about three hours.
- After arriving at his kidnapper's destination he heard chattering, some kind of talk he didn't understand. As it became light he saw there were four hairy beings, two big ones, a male and a female; and two adolescents — a boy and a girl.

- He was being held captive in about a 10-acre area bowl surrounded by mountains. On the southeast side there was a V-shaped opening.
- He had his rifle, but his box of rifle shells was missing from his sleeping bag. However, in his pocket he had 6 bullets for his 30-30.
- Ostman was held captive for six days. He thought he was about 25 miles from Toba Inlet, but wasn't sure. So he was trying to figure out which way to run once he got out of that giant bowl.
- His description of these beings was detailed with estimated heights, weights, hair coloring, facial appearance, jumping abilities, and so on.
- Ostman escaped by giving "the old man" a full box of snuff — "the old man" ate all of it and became sick. This allowed Ostman to escape.
- He came out at a logging camp on the Salmon Arm Branch of Sechelt Inlet. From there he caught the Union Boat to Vancouver.

Please note the Salmon Arm Branch of Sechelt Inlet is more than 60 miles from the headwaters of Toba Inlet. Very important: The Salmon Arm Branch is separated from Toba Inlet by a rugged mountain range, then Jervis Inlet, which along its lower reaches is more than a mile wide; and then, Zoonie Narrows, an inlet in a steep canyon. (See Google Earth)

On-site Reconnaissance

In June of 1992, Peter Byrne, Al Berry, Ken Corben and I decided to explore the details of Ostman's story. We flew to Campbell River, BC. Our plan was to fly over the area in my airplane, find the bowl and return to Campbell River. There, Peter had arranged to hire a helicopter to take us to the site where we would look for artifacts to add credibility to Ostman's amazing story.

After we arrived at Campbell River in my Cessna 210 we rested, fueled up and explored the Toba Inlet and surrounding area. After seven hours of flying low and slow, high and slow, we did not find the bowl. Plus, there was not a view of the inlet from 1,000 feet, not even 2,000 feet. So, we came to a different conclusion about the location of the story.

Headwaters of Tobas Inlet as seen from author's aircraft.
Note the rugged country beyond.
Confirm through Google Earth

Ostman's Account

Al Berry (pointing) and Peter Byrne near the headwaters of Toba Inlet where I landed on a sand bar.

Our time in Canada was running out, so we returned to the states and pondered. We assessed the complete story and all came to the conclusion that after 33 years Albert Ostman was mistaken on the inlet he was rowed to. After seeing the relationship of Lund, BC to the Toba Inlet, we also concluded it would have taken one hell of an old Indian to have rowed him that far in less than a day. (Ostman said they arrived at 4 p.m.).

Plus, as unyielding as we were — all looking desperately to find that bowl, we didn't find it. Our conclusion was that he was let off at one of the many other inlets prior to Toba.

Ostman's Account

Before making this idea public I decided to call John Green and run this thought by him. No matter how detailed I was about the research we did, he was adamant that it was Toba Inlet — that Ostman was absolutely truthful.

I told John that we were not questioning Ostman's honesty, but after 33 years he could have been mistaken about the details when interviewed. Especially since he was unclear (except for the captivity) about the accuracy of the other parts of the account. And that's how my version of the story ends. I'd like to add that its getting close to 100 years since Ostman's kidnapping took place; and, as much as it would be another fun adventure, any existing artifact still in the BC rain forest might be really difficult to locate.

Connecting Ostman to the Quantum World

Without gross speculation, it is not possible for me to connect the recorded experiences of Albert Ostman to the world of quantum physics. However, it is an enjoyable topic to reflect on. Indeed, you may see parallels to Ostman's story in preceding chapters of this book, chapters which may directly connect Bigfoot to quantum physics.

The conditions which may be speculative and connect to quantum physics are three: One; was he carried three hours, or more? (See Chapter 13, Quantum Time.) Two; when he woke, was it the succeeding morning, or another morning? And of course, Three; how far was he transported to the mountain citadel of Bigfoot? And the corollary questions — was Ostman taken directly to the citadel? Was he conscious the whole time?

Ostman's Account

Mouth of Tobas Inlet. Photo by Ron Morehead © 2016.

Finally, what was the real reason a hard-headed newspaper man, John Green, insisted that Albert Ostman was rowed to the head of Toba Inlet when a location in the lower Jervis Inlet would have provided Ostman with his 1,000 foot elevation view of bay and boats? Depending on the tide, that would make the rowing time believable. Was Mr. Green hinting at something in Ostman's story, something entirely speculative, something he did not want to print. Remember Al Berry's early warnings to me about credibility and keeping quiet about unusual events?

Afterword

> "Science must change, as it discovers that its net of evidence is equipped only to catch certain kinds of fish and that it is constructed of webs of assumptions that can only hold certain varieties of reality, while others escape its net entirely. Science, perhaps without being aware of it, has narrowed its field of vision." By Jesse Emspak, Live Science Contributor

We read from ancient texts, and surmise from Biblical scripture, that aliens have visited and played god with the DNA of a variety of earth species, including humans (Genesis 6:4). Call them what you will, these creatures, beings, peoples, angels, demons, things, sasquatches or bigfoots, exist on this earth. It is my belief that their probable nuDNA, from different alien sources could cause these beings to be of either an antagonistic or benevolent nature.

Dr. Ketchum was venomously attacked about her nuDNA findings. (See Chapter 12, Professional Findings). The results of her project would have enormous Judeo-Christian religious impact. Is the Bible right? Was there creation? Or is there evolution? Or, as I have suggested in this book, is there a blend of the two?

Did man evolve, and then God created His own hybrid, changing the troglodyte DNA, to match His own image of sapience? Shortly afterward, aliens descended with the intent of corrupting the human genome. Why? — they are envious of man (Psalms 8:4-6). We have what they want!

Because of my spiritual philosophy, in this book I have proposed that these beings may have different agendas — some to sway humans to "the dark side," and others to protect, and teach many of us how to return to our whole self, our pure self. It is for this reason that I hoped to express the importance of knowing just who

Afterword

you are as a human, whose image you were made in, and the filter of protection that we have been given if we so choose.

In Finality

Due to the mind boggling interactions that continue to happen at our *Sierra Camp* and beyond, I was led to quantum physics to figure this out. I would now like to reiterate the haunting puzzlements it has presented, quandaries that forced my inquisitive mind into 45 years of expensive, self-funded investigation and research.

I also note that while Bigfoot has many more observed attributes than in the following list, each item in this list has a connection to quantum physics per one or more chapters in this book.

1. I don't believe these beings share the same genome.
2. Some are antagonistic and some are benevolent.
3. Many may be a diluted hybrid from eons of inbreeding with Indigenous peoples.
   - Because of this some may have depleted their original abilities.
   - Many, if not all, are sapient—for sure I believe them all to be cognizant.
4. There is a strong argument that they are telepathic, via quantum entanglement.
5. Many have an incredible vocal range—exceeding mans'.
6. Many have their own rapid-fire language.
7. They can move inhumanly fast.
8. They may be able to change their vibrational frequency to be out of our perception.
   - They are interdimensional—changing from mass to energy, and back again.

Afterword

I truly hope this book will inspire the reader to delve into quantum physics; the correlations made herein, and the information I've presented. It can be a lot to digest if you've never considered the possibilities. But I believe the spiritual advancement of our species depends on embracing and practicing much of what is offered in these pages.

Ron Morehead, Sequim, Washington, 2017

Rhonda's Sighting

Although my daughter, Rhonda, told me about unusual events that had taken place with her around our Sierra Camp, she said she'd never seen the source that could have caused the events. It was in 1995, early in the evening, after she and I had eaten our meal, when she stood up from the log bench, turned around and a Bigfoot was looking at her.

She was speechless for a moment and could only point. By the time I'd asked her, "what?" she said "Bigfoot." I stood up and looked, but the Bigfoot had moved out of sight behind a huge tree — gone for the night, I imagined.

With the fading daylight, I thought it best to examine the area the next morning. I went to where Rhonda saw the creature while she stood where she was at the time of the sighing. She estimated that it was a little over 7-feet tall. She told me that it looked like a large black basketball player, but with very broad shoulders. As it turned, she said that is turned its whole body— without moving its neck.

Shortly afterward she and I looked for footprints but didn't find any—not unusual for the area around our camp, which has a lot of forest duff, being thick with pine needles.

Rhonda's Sighting

Artist: Sybilla Irwin

Friends of the Project

Molly Hart Lebherz: Editor

Don Cassity: Artist & Illustrator
http://donaldcassity.com

INDEX

"Awake and Empowered"
  Kevan Ryon 130
Ant Queen 109
Apauruseya 65
Araque, Rachelle Morehead iv
Aristotle 52
Baerbel (translator) 99
Bayanov, Dr. Dmitri 98
Berry, Al 55, 96, 203
  1976 Bigfoot book 204
Berry, Greg 212

Biblical
  1 Corinthians 13 x
  1 Corinthians 15:31 167
  1 Corinthians 6:19 130
  1 John 1:5 viii
  1 John 4 & 8 57
  1 Peter 2:9 23
  2 Chronicles 25:14 19
  2 Corinthians 6:16 130
  Ark of the Covenant 129
  Book of Enoch 18
  Cain 16
  Canaanites 18
  Corinthians 13:13 79
  Ecclesiastes 1:10 92
  Ephesians 1:4 69
  Ephesians 4:18 170
  Galatians 1:9 57
  Genesis 1:26 .... 21, 101
  Genesis 11:6 & 1-9 21
  Genesis 4:23 17
  Genesis 6:4 229

Hebrews 13:2 23
Hebrews ll:1 68
Jesus name is Yeshua ... 128
John 1:1 95
  1: 18 125
John 14:12 16
  14 128
  17 70
  9 125
John 17:21 .... 104, 129
John 3:3 69
John 8:12 .... 73, 125
Judkins, Dr. Aaron,
  archeologist 7
Luke 16:13 172
Luke 17:26 25
Luke 6:45 170
Maker communication
  with 140
Mammon 172
Matthew 16:19 70
Matthew 18:3 69
Matthew 21:53 129
Matthew 6:22 128
Nephilim 17, 193
Numbers 13:33-34 18
Proverbs 16:23 ix
Proverbs 18:21 95
Psalms 46:10 130
Psalms 8:4
  8:6   229
Rebekah: Esau
  Jacob 19
Tower of Babel 21

Bigfoot
- Inspect deer carcass..........12
- Jerry Crew names.............. 1
- Sapience in....................... 9

Bigfoot Attributes
- 13 foot betwen prints.......39
- 25 ½ inch prints 39
- Can BF disappear 49
- Cone cells
fovea centralis.......................33
- Did not turn neck 232
- Eyeshine and night vision 31
- Glowing eyes 34
- Glyphs...............................44
- Olfactory sense 35
- Sound graphs by Dr. Kirlin 37
- Sounds associated with ..36
- Speed of movement 35
- Tapetum lucidum
retina   31
- Telepathy..........................46
- Tree knocks 42
- Vocal ability 37, 47

Black budget science..........158
Black hole 83
Bludorf, Franz.......................99
- Vernetzte Intelligenz.....104
Bohr, Dr. Niels......................84
Bohr, Niels.............................56
Brain, human 169
Burr, Wendy 41
Burtsev, Dr. Igor...................98

Byrne, Peter.192, 225, 226, 241

Campbell, Keri....................191
Cannibal Giants
- Lovelock Cave, Nevada..27
- Paiute Indians 27
Cassity, Don
- Artist 3, 23
CERN 51
Clairvoyant children..........112
Closed timelike curves 156
Continental drift...................24
Cosmic War on Earth 23

Dark Energy..........................84
Dark Matter 83
Darwin, Charles 14
De Pretto, Olinto 197
Deep meditative state
- Brain waves....................134
Delayed Choice Experiment 160
Descartes, René...................128
Deutsch, Dr. David 156
Dispenza, Dr. Joe
- "Evolve your Brain" 137
DMT
- Di-methyl-tryptamine
Psychedelic compound 132
DMT space
- Beings encountered.......135
DNA..............................99, 105
- Genbank 152
- mtDNA & nuDNA 152
- nuDNA 229
Dr. Michio Kaku...................52

Dyke, Professor Daniel J.
    New earthly timeline 25

Earth-shattering event
    About 12,000 BC 25
Edison, Thomas 56
Edwards, Guy
    Brainjar Media iv
Ego 167
Einstein, Albert . viii, 9, 53, 63, 76, 197
    "Laughter of the gods." 127
    Creative Force 200
    E = mc2 197
    E=mc2 125
    Feldwebel 198
    Spooky action at a distance 103

Fahrenbach, Dr. W. Henner
    Primate expert 31
Feynman, Dr. Richard 157
Fluoride 136
Foerster, Brian
    Ancient SA culture expert 186
Fosar, Grazyna 99

Garjajev, Pjotr 99, 106
Giants
    Peru and Equador 1
Gigantopithecus
    Ralph von Roenigswald found 5
Gimlin, Bob i
God Particle 51

Gosar, Grazyna
    Vernetzte Intelligenz 104
Grandfather paradox 158
Green, John
    Sassquatch, The Apes Among Us 222

Haroche, Dr. Serge 161
Hawking, Stephen 64
Heart, human
    Field generator 171
Henry, Dr. Hugh
    New earthly timeline 25
Hermanns, William 198
Higgs Boson Particle 51
Hilton, Donald 40
Hinduism 65, 176
    Rig-Veda 126
    Upanishads 176
    Yogatattva 176
Hoax 203
Human sound expert
    Nancy Logan 143
Hybrid
    Common cranial elongation 7
    Dr. Stuart Fleischmann 6
    Pharaoh Akhenaten possible 6
Hybrids 15
Hyper communication 111

Invisibility
    Aborigines 177
    Bigfoot 180
    Clackamas Indians 180

Cloak Technology..........174
Dege, Erik......................180
Hermetic Order..............178
Higbee, Donna................175
Human Involuntary......175
Native American...........180
Theosophical Society.....178
Irwin, Sybilla artist.............233

Johnson, Lewis......................11
Johnson, Ronika Morehead. iv

Kaku, Dr. Michio..................52
Kant
    Critique of Pure Reason199
Kassewitz, Jack......................93
Kern Peak,
    Bigfoot track...................184
Ketchum, Dr. Melba...........229
    Sasquatch DNA analysis
   152
Kirlin, Dr. Lynn
    Bigfoot sound graphs......37
Kirlin, Dr. R. Lynn.................96
    Curriculum Vitae...........142
    Photograph......................141
Krantz, Dr. Grover................59

Laser light.............................110
Lawrence Livermore
    National Laboratory........48
Light, speed of.......................71
Logan, Nancy
    Human Sound Expert...143

MACHOs...............................91

Manlike Monsters on Trial
    by Dr. R. Lynn Kirlin....142
Masons
    John Macky.....................177
McDowell, Bill................11, 40
Megalith Walls
    Sacsayhuaman, Peru.....185
Melatonin
    Pineal Gland...................131
Moody, Richard Jr..............197
Morehead O'Connell, Rhonda
   41
Morehead, Royce................. iv

NASA Goddard Space Flight
    Center................................45
NASA Science
    Ann Field..........................85
Nelson, R. Scott......41, 94, 119
    Boring, J.E. Vetting........149
    Crypto-Linguist.............148
Nepal.....................................241
Neuroplasticity...................168
Newton, Sir Isaac.........53, 197

Ocean, Joan...................93, 213
    Kiamichi Mountains.....216
    Participatory Research..213
    Tel-Empathic
        Communication........213
    Telepathic
        communications.......217
O'Connell, Rhonda Morehead
   7
    Sighting...........................232
Ostman, Albert...................221

On-site reconnaissance . 225
Police magistrate 222
Quantum world
 connection 227
Salmon Arm Branch 224
Toba Inlet 223

Pangaea Supercontinent 24
Participatory research
 Dolphins 215
Paulides, David
 Missing 411 20
Pendry, Dr. John 174
Penrose, Sir Roger 61
Peres, Asher
 Early quantum theorist 160
Peru
 2014 Expedition 185
Peruvian skull
 naturally elongated 8
Pineal gland 69
 Calcified in most people
  136
 Melatonin 131
Pineal Gland
 Ghaga benefits to 138
 Sun-gazing 139
 The third eye 125
Planck, Dr. Max 62, 167
Planck, Max 53
Plato 52
Poincaré, Jules Henri ... 72, 197
Poponin, Dr. Vladimir 99
Portals 45
 Ocean, Joan 219
 X points 46

Powell, Dr. Cynthia
 Animal night vision 34
Pre-Inca Culture 186
Pre-Incan culture 182
Preskill, Dr. John 83
Preston, S. Tolver 197
Psychedelic compound
 Pineal gland production of
  132
Puma Punku, Bolivia
 Lake Titicaca 26
Pye, Lloyd
 Star Child 188

Quantum Entanglement
 ... at its Best 117

Raja yoga 176
Ralph, Dr. Timothy 157
Ringbauer, Martin 158
Rosicrucianism 177
 H. Spencer Lewis 177
Rozencwaig, Dr. Roman
 Pineal and melatonin ... 131
Russian DNA Research **99**

Sanskrit 65
Sasquatch people 103
Schumann resonance
 Schumann, Winfried O. 120
Science's Microscope 57
Siddhis 176
Sierra Camp .... x, 203, 230
Sierra Camp shelter 13
Skull, naturally elongated
 Peruvian Paracas infant 189

Sleep
  Waves and Frequencies 121
Sound to Light 48
Star Child
  Lloyd Pye........................188
State Darwin Museum of
  Science..............................98
Stevenson, Dr. Ian 67
Strassman, Dr. Rick
  DMT The Spirit Molecule
 133
Syntonic Research, Inc. 96

Tapes, Bigfoot 205
Tartini, Giuseppe.................109
Teibel, I. E.
  Syntonic Research 205
Teibel, I.E..............................96
  Mike Kron, associate 208
Temple 129
Tesla, Nikola 65, 75, 173
The Oregon Bigfoot Highway
  Willamette City Press LLC
 180
Thorne, Dr. Kip.....................83
Time
  and Light Speed.............164
  Inseparable from space .163
  Is fourth dimension 162
Time travel
  Dr. Hawking on 155
Troglodyte..........................229

Type I civilizations.............113
Type II civilization 113

UFO......................................183
University of Queensland
  Photon time travel.........155
Ussher, Bishop James
  Timeline of Creation 25

Vacuum domains 114
Vedas 65
Vernetzte Intelligenz ...99, 104
  "Networked Intelligence"
 104
Vibrational Frequency
  Earth...............................120
  Human............................120
Voices in the Wilderness
  by Ron Morehead.............. v

Wattles, Wallace 78
WIMPS..................................91
Wineland, Dr. David 161
Wood knock........................119
Wormholes..........................114

Yang Hao
  Professor QMUL............174
Yeti 98
Yoga sutra
  Patanjali, author 176

Nepal

The Himalayan Mountains at sunset from Pokhara, Nepal.

2011: For decades Peter Byrne has been a legend in Nepal, home of the legendary Yeti. He and I spent three weeks exploring the jungles of western Nepal assessing the tiger and leopard population.

Nyatapola Temple in Bhaktapur, Nepal. Erected 1701 – 1702.

Peter Byrne (driving) and the author enroute to adventure.

Author at the Shuklaphanta Reserve, western Nepal

Products

"Bigfoot Recordings Vol. 1"

A 40 minute CD (below), features very clear Bigfoot vocalizations captured by Al Berry, the investigative reporter, who was invited into the Sierra Camp. Narrated by Johnathan Frakes, Star Trek, The Next Generation.

"Voices in the Wilderness"

above is Ron Morehead's chronicle of his four-decade adventure at the Sierra Camp, and the interactions with a group of Bigfoot. Included is a CD with actual interactive vocalizations.

"The Quantum Bigfoot"

Featured below, is Ron's second book and delves into the science that explains the enigmas often associated with these beings.

"Bigfoot Recordings Vol. 2"

Pictured above, is a 40-minute CD, produced and narrated by Ron Morehead. Two years after Vol 1, at the same remote camp. Features vocal interaction between Ron and these beings.

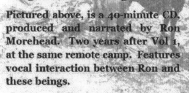

www.http://ronmorehead.com/books-audio/